PROTOTYPE NATION

T0219453

Princeton Studies in Culture and Technology

Tom Boellstorff and Bill Maurer, series editors

This series presents innovative work that extends classic ethnographic methods and questions into areas of pressing interest in technology and economics. It explores the varied ways new technologies combine with older technologies and cultural understandings to shape novel forms of subjectivity, embodiment, knowledge, place, and community. By doing so, the series demonstrates the relevance of anthropological inquiry to emerging forms of digital culture in the broadest sense.

Prototype Nation

China and the Contested Promise of Innovation

Silvia M. Lindtner

PRINCETON UNIVERSITY PRESS

PRINCETON AND OXFORD

Published by Princeton University Press
41 William Street, Princeton, New Jersey 08540
6 Oxford Street, Woodstock, Oxfordshire OX20 1TR

press.princeton.edu

All Rights Reserved
ISBN 9780691179483
ISBN (pbk.) 9780691207674
ISBN (e-book) 9780691204956

British Library Cataloging-in-Publication Data is available

Editorial: Fred Appel and Jenny Tan
Production Editorial: Sara Lerner
Jacket Design: Pamela L. Schnitter
Production: Brigid Ackerman
Publicity: Kate Hensley and Kathryn Stevens
Copyeditor: Karen Verde

Cover Credit: Courtesy of Cao Fei, Vitamin Creative Space and Sprüth Magers

This book has been composed in Adobe Text and Gotham

Printed and bound by CPI Group (UK) Ltd, Croydon, CR0 4YY

for my mother

Prototype, n. / ˈprəʊtətʌɪp/ 1. A first or preliminary version of a device or vehicle from which other forms are developed. 1.a. The first, original, or typical form of something; an archetype.

—*OXFORD ENGLISH DICTIONARY*[1]

CONTENTS

LIST OF ILLUSTRATIONS

ACKNOWLEDGMENTS

This book is born out of many personal and professional friendships that have shaped my life. First and foremost, the three people without whom none of this would have been possible are my mother Eva-Maria, whose belief in me and whose fight for justice and equality for women and people of color have provided the strength to divert from the familiar and comfortable—from her accomplishments as a single mother to her achievements in a male-dominated engineering industry, she is a role model for me; my partner William, who has walked by my side through so many adventures and whose relentless care has supported me in following the paths that demand courage and persistence; my PhD advisor Paul Dourish, whose intellectual brilliance and generosity continue to guide me in my scholarly pursuits—I could not have accomplished this project without his inspiring wisdom, feminist commitments to making our sociotechnical worlds otherwise, and his companionship during field research in China, analysis, writing, and so much more.

I am indebted to the many people who let me into their lives and spaces of work while I conducted the research that informs this book. Many of the people who have deeply influenced my thinking—especially many of the Chinese women who labored for the international innovation industries—I cannot properly thank here for reasons of anonymity and protection.

I am deeply grateful to aaajiao, David Li, Eric Pan, Chen Xu, and Liu Yan for their friendship, for many years of collaboration, and for so generously sharing their wisdom and time. I am equally indebted to Jeffrey Bardzell, Shaowen Bardzell, Anna Greenspan, and Ingrid Fischer-Schreiber—for their friendship and for traveling with me through intellectual terrain and field research. Our collaboration for research and writing, and their constructive critiques of various drafts, have centrally shaped this book. I also owe a debt of gratitude to Daniela Rosner for the close reading and commenting on numerous (including very early and very drafty) versions of this book, and to Anna Watkins Fisher for the countless hours of brainstorming and for the stimulating conversations about the book's key concepts.

Nancy Chen's and Fred Turner's insightful and meticulous comments they provided as external readers during my manuscript workshop in March 2018 were an enormous help to me. Their close reading of the draft has helped sharpen key analytical concepts and the book's contributions. The manuscript workshop itself would have been impossible without Paul Conway's brilliant work as workshop chair, Paul Edward's tremendous support leading up to the workshop, and Mary Gallagher's and Elizabeth Yakel's generous efforts to provide the resources that enabled an interdisciplinary and constructive workshop environment. I thank the manuscript workshop attendees for their thoughtful feedback: Mark Ackerman, Seyram Avle, Iván Chaar-López, Maura Cunningham, Jerry Davis, Mary Gallagher, Cindy Lin Kaiying, Jean Hardy, Anna Watkins Fisher, Lisa Nakamura, Joyojeet Pal, and Christian Sandvig. Ken Shirriff was so kind to read and comment on a full draft of the book. I am immensely grateful for his detailed engagement with the language and key analytical concepts of the text.

Kim Greenwell, Laura Portwood-Stacer, and Heath Sledge have provided invaluable input during different stages of the developmental editing process. I thank Laura Portwood-Stacer for her careful read of and sharp commentary on a full draft in 2018. I am grateful for Kim Greenwell's rigorous edits of the introduction to help translate scholarly commitments into an approachable and concise read. And I can't thank Heath Sledge enough for her laborious engagement with the whole manuscript, crucial in helping me refine my arguments and pull together my ethnographic stories into a compelling whole. I thank my coach Rena Seltzer for guiding me with wit, wisdom, and care to regain energy and focus in the darkest moments of analysis, writing, and faculty life.

At the University of Michigan, I am most fortunate to have received close mentorship and to have learned from Mark Ackerman, Jerry Davis, Paul Edwards, Nicole Ellison, Gabrielle Hecht, John King, Lisa Nakamura, Sally Oey, Christian Sandvig, and Elizabeth Yakel. I have tremendously benefited from conversations and/or collaborations with Robert Adams, Megan Ankerson, Irina Aristarkhova, John Carson, Yan Chen, Paul Conway, Tawanna Dillahunt, Ron Eglash, Ana Sabau Fernandez, Anna Watkins Fisher, Mary Gallagher, Patricia Garcia, Matthew Hull, Margaret Hedstrom, Julie Hui, Pedja Klasnja, Meena Krishnamurthy, Rebecca Modrak, Sarah Murray, Guna Nadarajan, Lisa Nakamura, Joyojeet Pal, Rebecca Pagels, Shobita Parthasarathy, Casey Pierce, Aswin Punathambekar, Sarita Yardi Schoenebeck, Perrin Selcer, Andrea Thomer, Nick Tobier, William Thomson, Kentaro Toyama, Antoine Traisnel, Oscar Ybarra, Tiffany Veinot, Zheng Wang, among others. I learned so much from John Carson and Gabrielle Hecht, especially while we taught together in the UM STS program. Anna Watkins Fisher, Sarah Murray, and Antoine Traisnel deserve special appreciation for years of companionship and supportive co-presence in collocated writing on and off campus. Irina Aristarkhova and Guna Nadarajan have provided tremendous intellectual and emotional support throughout the years. I am grateful to Thomas Finholt, dean of UMSI, as well as James Holloway and Amy Conger, who have shaped UM's global engagements initiatives in the provost office, for their generous support of my research in China and beyond. I thank the various members and collaborators of Precarity Lab—Cassius Adair, Irina Aristarkhova, Anna Watkins Fisher, Iván Chaar-López, Meryem Kamil, Cindy Lin Kaiying, Lisa Nakamura, Cengiz Salman, Tung-Hui Hu, Kalindi Vora, Jackie Wang, and McKenzie Wark—for sharing a collaborative space of interdisciplinary thinking and writing. Lisbeth Fuisz and Alan Klima have provided tireless encouragement and writing tips in the writing boot camp Academic Muse.

I thank my doctoral students and postdocs, past and present, for their intellectual companionship and friendship. Their commitments to think and make the world otherwise give me strength to persist in standing up against injustice at the university and beyond. I learn so much from working with them and guiding them through various stages of their professional lives. During the period of researching and writing this book, this included Seyram Avle, Jean Hardy, Cindy Lin Kaiying, and Stefanie Wuschitz. I have also benefited from working closely with students who signed up for independent study with me, who became fellow researchers and writers, or on whose doctoral committees I had the privilege to serve; Padma Chirumamilla, Meryem Kamil, Liz Kaziunas, Iván Chaar-López, Victoria Koski-Karell, Jonathan Riley, and Cengiz Salman.

I cannot express enough gratitude for the long-time friendship, intellectual exchanges and input, and emotional support provided by Julka Almquist, Marisa Cohn, and Katie Pine throughout and well beyond graduate school. They were the ones who made me feel that graduate school was a place where I belonged. I am so fortunate for their unconditional belief in me as a person and as a scholar. I deeply treasure the memories of our shared times, too numerous to recount here in detail but traveling with me, always. During my time at UC Irvine, I have learned much from the members of my dissertation committee Tom Boellstorff, Mimi Ito, Melissa Mazmanian, and Jeffrey Wasserstrom as well as from Yunan Chen, Beatriz da Costa, Martha Feldman, Susan Greenhalgh, Gillian Hayes, Antoinette LaFarge, Roberta Lamb, George Marcus, Keith Murphy, Bonnie Nardi, Robert Nideffer, Simon Penny, Kavita Philip, David Redmiles. Fellow doctoral students at the time, at and beyond UC Irvine, have been a resource of past and ongoing support: Mary Amasia, Morgan Ames, Arianna Bassoli, Eric Baumer, Johanna Brewer, Kenneth Cameron, Judy Chen, Maura Cunningham, Marc DaCosta, Xianghua (Sharon) Ding, Lynn Dymbrovski, Madeleine Clare Elish, Megan Finn, Ellie Harmon, Garnet Hertz, Lilly Irani, Lucian Leahu, Hrönn Brynjarsdóttir Holmer, Paul Morgan, Bruno Nadeau, Josef Nguyen, Lilly Nguyen, Elisa Oreglia, Sun Young Park, Daniela Rosner, Luv Sharma, Christo Sims, Marcella Szablewicz, Amanda Williams, among others. Charlotte Lee, Irina Shklovski, and Janet Vertesi during their time as postdoctoral fellows at UC Irvine have taught me much, from teaching techniques to positioning myself in the interdisciplinary worlds of critical technology research. I deeply enjoyed and benefited from Garnet Hertz's and Amelia Guimarin's companionship during phases of research on DIY making and hacking in the United States and China. Madeleine Clare Elish has been a true friend throughout so many life stages since the dissertation. Without her humor and wit, academia would be a much darker place. I have learned from and am deeply grateful for the friendship of Marc DaCosta as we navigated the often overlapping worlds of academia, entrepreneurship, and investment.

I owe special thanks to Denisa Kera and Tricia Wang, my fellow travelers through the worlds of technology and China research during and after graduate school. Tricia Wang is an ongoing inspiration for how to connect feminist and critical commitments to worlds outside the academy; her intellectual support and emotional presence during various stages of fieldwork in China were tremendous. At a conference on "Governing Futures" hosted by the STS program at the University of Vienna in 2011, numerous people kept insisting that I must "talk to" Denisa Kera, who had attended the event; they were right—that day in Vienna was the beginning of a long friendship and many collaborations that took us from Austria to China and many other places, for which I am so deeply grateful. Part of fieldwork in China was also centrally shaped by the generous support I received from the inspiring researchers at Intel I had the privilege to work with since 2008: Ken Anderson, Maria Bezaitis, Scott Mainwaring, and Anne McClard. I owe special thanks to Ken Anderson, who has taught me how to be an ethnographer. I am also deeply indebted to Monroe Price for his continuous support of me and my work since we met at a conference at USC during my third year in graduate school. He also introduced me to the brilliant Briar Smith, who I owe much for her crucial support in making possible a workshop in Budapest in 2011 that would become the first of many interdisciplinary gatherings I helped organize between makers, scholars, policy makers, industry, and activists over the years to come. Julie Starr offered me her friendship and provided tremendous support during the isolating phases of dissertation writing in Shanghai. I feel privileged to have been able to learn about Shenzhen from Mary Ann O'Donnell. I thank her for many hours of stimulating exchange, the most remarkable tours of the city, and profound insights into the region's histories and presents. Lyn Jeffery has been a wonderful co-traveler to the international worlds of tech innovation between Shenzhen and Silicon Valley, who has guided ethnographic paths and analytical sensibilities.

I have received insightful comments and constructive feedback on various stages of writing for this book from a stunning array of people for whose generosity, mentorship, example, and critique I am overwhelmingly grateful. These include Irina Aristarkhova, Jonathan Bach, Padma Chirumamilla, Maura Cunningham, Vishnupriya Das, Paul Edwards, Kirsten Foot, Victoria Hattam,

Christina Dunbar-Hester, Anna Watkins Fisher, Mark Fraser, Anna Greenspan, Phil Howard, Karin Knorr Cetina, Aynne Kokas, Victoria Koski-Karell, Jochen Metzger, Michelle Murphy, Sarah Murray, Gina Neff, Mary Ann O'Donnell, Katy Pearce, Daniela Rosner, Perrin Selcer, Ken Shirriff, Alex Taylor, Antoine Traisnel, Cara Wallis, and Daniel Williford. At workshops, at conferences, symposia, and lectures where I have presented my research that informs this book, I had the good fortune to discuss aspects of my thinking and to receive invaluable insights from Sareeta Amrute, Liam Bannon, Mario Biagioli, Nicola Bidwell, Matthew Bietz, Pernille Bjørn, danah boyd, Marisa Brandt, Margot Brereton, Anita Say Chan, Lily Chumley, Gabriella Coleman, Elli Blevis, Alessandro Delfanti, Jill Dimond, Carl DiSalvo, Yige Dong, Pelle Ehn, Madeleine Clare Elish, Ulrike Felt, Geraldine Fitzpatrick, Laura Forlano, Sarah Fox, Christopher Frauenberger, Verena Fuchsberger, Liz Gerber, Melissa Gregg, David Hakken, Brian Hartman, Victoria Hattam, Jo Havermann, Marie Hiecks, Kia Höök, Tom Igoe, Steve Jackson, Wendy Ju, Naja Holten Møller, Lone Koefoed Hansen, Aynne Kokas, Mike Kuniavsky, Liz Lawley, Eugenia Lean, Débora Lenzeni, Youn-kyung Lim, Ann Light, Laura Liu, Shannon Mattern, Annette Markham, Martina Muzi, Nicole Nelson, Josef Nguyen, Eric Paulos, Nadya Peek, Marianne Petite, Irene Posch, Allison Powell, David Ribes, Louisa Schein, Victor Seow, Phoebe Sengers, Bart Simon, Jack Qiu, Matt Ratto, Janet Roitman, Sophia Roosth, Natasha Schull, John Seely Brown, Clay Shirky, Rachel Smith, Johan Söderberg, Stephanie Steinhardt, Luke Stark, Lucy Suchman, Karen Tanenbaum, Theresa Jean (Tess) Tanenbaum, Manfred Tscheligi, Austin Toombs, Guobin Yang, Fan Yang, Anna Valgårda, Judy Wajcman, Ron Wakkary, Patrick Whitney, Lu Zhang, and Caitlin Zaloom.

The completion of this manuscript has benefited from the generous support from the Lieberthal-Rogel Center for Chinese Studies, the School of Information, and the Advance Program at the University of Michigan. The research that has informed this book was in part funded by research grants from the National Science Foundation, the Institute of Museum and Library Services, a Google Anita Borg fellowship, a Fudan University post-doctoral fellowship as well as funding from Intel Labs and the Intel Center for Social Computing at UC Irvine. I thank Deborah Apsley, Stacy Callahan, Jacques Chestnut, Jocelyn Jacobs, Nayiri Mullinix, Rebecca O'Brien, Todd Stuart, Heidi Skrzypek, and Jocelyn Webber for their tireless administrative support at the University of Michigan.

I express my deep gratitude to friends who have accompanied me through various stages of research, writing, and professional development: Adam Brooks, Matthew Burgess, Nunzia Carbone, Susan Dai, Rick Hollander, Witold Klajnert (a.k.a. Fossibaer), Katharina Kuebrich, Lisa Juen, Julian Juen (a.k.a. Seniorbaer), Chris O'Brien, Barbara Ratzenbock, Mike Romatowski, Monica Shen, Julie Starr, and Francesca Tarocco.

I can't thank enough the two anonymous reviewers who read various stages of this book from proposal to final manuscript. They provided tremendously constructive and sharp criticism, which centrally shaped my thinking and writing. With Fred Appel, Princeton University Press has a committed and engaged executive editor who I have tremendously enjoyed working with and who I thank for his support throughout the process. I am also deeply grateful to the PUP staff, especially Sara Lerner, Theresa Liu, and Laurie Schlesinger, as well as Karen Verde, for their wonderful support in bringing this book into existence. And last but certainly not least, I am indebted to Tom Boellstorff, who has worked with me on various stages of this book from a close mentor in graduate school to his role as special series editor with Bill Maurer of "Culture and Technology" at Princeton University Press.

PROTOTYPE NATION

1

Introduction

THE PROMISE OF MAKING

Promises . . . point us somewhere, which is the where from which we expect so much. . . . The promise is also an expression of desire; for something to be promising is an indication of something favorable to come.

—SARA AHMED, *THE PROMISE OF HAPPINESS*, PP. 29-30

What is a prototype? The term prototype is typically used in the context of industrial production, design, and engineering; a prototype is built to model or test demand (from investors or users) for an idea or a product. But can we speak of prototyping a city, a region, a nation, or new ways of being? What would these complex prototypes look like, and what would they do? This book tells the story of how prototyping at vast scales came to be viewed as a promising way to intervene in entrenched structures of inequality, exploitation, and injustice—and how this promise became a demand for individual self-upgrade and economic development. As an ethnographer, I spent ten years (2008–2018) following the people who came together around the idea that cities, regions, economies, and even nations and life itself can be prototyped. They argued that if the production of technology was made available to everyone, concrete alternatives to corporatized, exploitative, and politicized technology could be tested. They envisioned that if people became makers of technology, they would own the things they made and could decide for themselves what their technologies—and by extension their social, economic, and political lives—would be like. The prototypes of intervention they made came to be widely known as the "maker movement."

1

This promise of making—that *every* individual can prototype and thus intervene at scale—was fundamentally exhilarating. It felt empowering to many, like a moral form of hacking; an ethical, democratized technological resistance that was experimenting with how technology can be otherwise. This book unpacks in ethnographic and historical detail how this happened; how "making" became saturated with an *affect of intervention*—a feeling of agency and control, a sense that alternatives to dominant structures at various spatial and temporal scales were possible. This affect of intervention created seemingly shared visions for the future—even when those visions were incompatible and contradictory. Making was taken up simultaneously to articulate a return to "made in America"—as former US president Barack Obama had envisioned it in 2013—and to overcome "made in China" and its associations of China with backwardness, low quality, and fakery. It was taken up by people, institutions, and corporations that we would typically think of as holding sharply opposing views; feminist technology researchers and designers, venture capitalists, educators, major tech corporations from Intel to Tencent, designers, technology activists, major governments with opposing political views, critical scholars of science and technology. The uptake of making was driven simultaneously by desires to relive modernist ideals of technological progress *and* by projects aimed at relocating future making and decolonizing technology and design. It was articulated both in terms of a nostalgic longing for older, "better" times *and* as a toolkit to imagine alternative futures. It became a site to rearticulate the importance of craftsmanship and its associations with individual self-transformation and autonomy. At the same time, it became a resource to envision an alternative designer, engineer, and computing subjectivity that challenged ideals of the autonomous self. The interesting question is not which version is true, but how it was possible for making to be understood through such contradictory terms.

I use the prototype as both an analytical concept and an emic term, i.e., as practiced in technology production and design. As anthropologist Lucy Suchman and colleagues note, the prototype has "particular performative characteristics within the work of new technology design."[1] It is a material and concrete proposal of alternative ways of thinking about technology and its role in the world, not "simply as a matter of talk, but as a means for trying the proposal out." In other words, the affective qualities of the prototype lie in its simultaneous functioning as object (a model) and process (testing). The term refers to both the normative modeling—the making concrete, or realizing—of specific ideas *and* the making of an alternative, which carries the potential for contestation and intervention. One of the key promises of the maker movement was that prototyping—the testing and modeling of a technological alternative—was no longer reserved for elites, for scientists, designers, or engineers. Rather, the techniques of "making" from reverse engineering closed systems to building your own devices and machines with open source hardware platforms and tools would make prototyping (and thus the testing and modeling of alternatives) available to everyone. A flurry of maker and open source hardware prototypes made this promise of making concrete; the "DIY cellphone" showed that *you*—rather than a big corporate player like Apple—can control the design and inner workings of your communication devices;[2] the hacking of proprietary health devices demonstrated how *you* can regain ownership of your body's data;[3] open source 3D printers made palpable how *you* can mass-produce in your own home.[4] All of these projects functioned as *prototypes of intervention*. They "demo-ed" how to see oneself as capable of intervening in technological ownership, industrial production, economies of scale, broken healthcare systems, and of undoing established notions of the good life. They modeled how to see oneself as in control of what technologies—and by extension one's social, economic, and political life—could look like. They created a feeling of being able to intervene at scale, from the individual body to the nation.

I use the prototype as an analytical concept to attend to a broadening disillusionment with digital technology and the IT industry. This book shows that ideals and practices of making spread in the very moment as the political and economic regime of techno-solutionism, i.e., the construal of complex social and economic inequities as problems that can be solved by technological solutions, began to be more widely critiqued. Making became more prominent during a time when people began reckoning with the tech industry's complicity in enabling structures and processes of exploitation, racism, sexism, and exclusion. It was a moment of realization that the structures and processes of capitalism had never been "external" to or "above" the workings of technology and design. The historical condition that gave rise to making was marked by a coming to terms with how technology had enabled the entrenchment of what is commonly thought of as key characteristics of neoliberal capitalism: the economization of the environment, of natural resources, and of life itself in the name of progress and development; the demand placed on individuals to self-actualize as economic agents made responsible for their own survival; the displacement of people and animals in the name of national sovereignty, global competitiveness, and security.

Making's particular local and translocal formations unfolded through a growing distrust of some of the basic assumptions of modernity itself. It emerged through and alongside a (belated) realization by members of tech and design industries and research that the promise of modern, technological progress and techno-solutionism had occluded and thus legitimized the violence and loss caused by capital accumulation and economic development. Advocates of making were less interested in finding a technological fix than in redefining what technology or a technological solution meant in the first place. They were invested in experimenting with alternative ways of conceiving of and producing technology, which ultimately recuperated the promise of happiness[5] and the good life attached to technological progress, precisely as people realized that these feelings had, in fact, long been unattainable for most.[6] Making, in other words, was simultaneously an expression and refutation of technological promise.

To make sense of this seeming contradiction, this book offers a genealogical approach that attends to the displacements of technological promise. It examines how technological promise can coexist with the proliferating distrust of its attainability. I show that the endurance of technological promise works through its displacement to sites formerly conceived of as the tech periphery, once portrayed as incapable of innovating and "in need" of technological intervention and economic development. Specifically, my focus is on China and how its image began shifting in the broader tech imagination at the very moment that the promise of making took shape and modernist ideals of technological progress were more broadly challenged.[7] I show how China, and more specifically the city of Shenzhen in China's Southeastern province of Guangdong, alongside other regions—including regions in the postcolonies, regions in rural America, former manufacturing cities in Europe and the United States—were rearticulated by a range of actors in the global tech industry, investment, policy, and politics as places where the future was now made, as newly innovative exactly because they were considered to be backward and thus not tainted by capitalism or modernity the same way.[8] These displacements of technological promise co-produced China as a *prototype nation*.

By *prototype nation*, I mean the stipulation that a nation can function as a prototype—a nation that can serve as the raw material for a new model (for instance, an alternative to established models of modern progress) or can generate demand for a particular kind of future (for instance, a nation's future freed from past and ongoing colonialism). The idea that a region, even at the scale of the nation, can function as a prototype, as a means of modeling a new way of life for others, and as an archetype that makes certain futures felt, concrete, and "masters the unknowable," is of course not new; it is historically constituted through projects of modernization, economic development, and colonization.[9] The European invention of the "nation-state," the historian Arif Dirlik reminds us, "was the ultimate vehicle of modernity."[10] The European nation was positioned as the prototype of modern progress and economic development, positing Europe as the archetype and "first" that had let go of feudal pasts and traditions. Other regions were construed as stuck in the past and as "in need" to learn from the Western "model" nations. Postcolonial studies have shown in great depth how the making of the Western nation as *the prototype* of modernization and development was contingent on the invention of "the Third World" as "other"—the discursive construct of a less civilized "other" that legitimized the extraction of resources the West needed for its own project of progress and economic development.[11]

The colonial project of prototyping a certain way of life (and demanding that others model themselves after that particular image) endures through the ideals and practices of technology innovation. It lives on in the construal of Silicon Valley's methods, instruments, and ideas of technology design and engineering as universally applicable.[12] And it is sustained through projects aimed at replicating Silicon Valley's "regional advantage"[13] elsewhere—from efforts to build the Silicon Valley of Russia in Skolkovo, a suburb of Moscow, to claims of the emergence of a global creative class.[14] It is most recently reactivated through the displacement of technological promise and the stipulation that "regional advantage" is now located elsewhere (in China, in rural America, in sub-Saharan Africa, etc.), at new frontiers, producing new horizons of possibility and investment opportunities—it is these displacements of technological promise that centrally concern this book.

I offer *displacements of technological promise* to bring into focus the violence and loss that are produced and yet often occluded by the endurance of technological dreams of future making. Weaving together sensibilities from feminist anthropology, critical race studies, and science and technology studies, this ethnography shows that the displacement of technological promise onto what was once imagined as the periphery of technological future making is a discursive move with material consequences, providing legitimacy for the reordering and restructuring of space and people, the flow of investments into certain spaces and technology practices rather than others, the casting of certain people as deserving while continuously keeping others on hold, framed as not (quite) ready, not capable of their own self-investment. Displacements of technological promise are *not* a linear movement of technological ideals and objects from "here" (the so-called developed world) to "there" (the so-called other part of the world). As I will show in this book, they unfold, instead, through circular, recursive moves, the recuperation of certain pasts and the silencing of others. They require labor and active maintenance. They thrive on the inclusion and exclusion of select sites and bodies. Displacement, anthropologist Juno Salazar Parreñas theorizes, is the "slow violence" that "works over multiple scales and beyond the clean boundaries of specific events, places, and bodies affected."[15] The displacements of technological promise I document in this book are not the same and yet are not unlike the slow violence Parreñas observes as materially experienced through eviction, mega-dam construction, and natural resource extraction. My particular focus is on how displacements of technological promise imbued neoliberal projects of regional laboratories, special economic zones, and smart city planning with a renewed promise of happiness, despite their histories of extraction, displacement, and violence.[16] When technological promise is at last granted to places and people that long yearned to be seen as just as innovative and creative as places like Silicon Valley, acts of violence and control in the name of innovation become less noticeable, occluded by the promise of modern technological progress and its associations with the good life[17]—the promise to be at last freed from colonial and racial "othering."

The idea that China constitutes a place to prototype alternatives to existing models of modern technological progress would have sounded absurd to most people ten years ago, when I started the research that informs this book. This is less so more recently. Western China commentators and news media have variously proliferated a sense that we are witnessing the rise of an emboldened China—one that more forcefully and assertively demands that the world, and the West in particular, ought to take it seriously as an equal (or threatening) player in the global political economy and in technology innovation in particular. Indeed, some speculate that the twenty-first century will be China's century. This book complicates these narratives of what historian Gabrielle Hecht calls "rupture-talk."[18] The notion of rupture cannot explain China's current moment, nor the ideals of the maker movement; indeed, "rupture talk" renders invisible the contingencies between past and present dreams of China as an alternative "model" or *prototype nation*, the rise and spread of the promise of making, and displacements of technological promise.

This ethnography is attuned to the genealogies of technological promise; it attends to the occluded contingencies of colonial pasts that remoralize "neoliberal exceptions"[19] and the endurance of technological promise in the present. Its analysis folds through recuperations and reappropriations, rather than "tracing" global flows and historical continuity. One of the most pressing tasks today, anthropologist Ann Laura Stoler urges, is to examine how imperial formations are refashioned, "often opaque and oblique," seemingly indiscernible, and escaping scrutiny.[20] This book attends to such "colonial reverberations" in the tech industry and digital technology projects whose promises of participation, peer production, and entrepreneurial agency occlude their historical contingencies.

In the following sections I provide a cursory sketch of the spatial and temporal contingencies of technological promise that this book attends to at length. I turn to 2015, when a high-profile Chinese politician and a well-known figure of the American maker movement each articulated a vision of making that was—without any explicit reference to the other—aligned in seemingly paradoxical ways that would escape our view if we restricted our analysis to the boundary of the nation or to an ahistorical approach.

The Socialist Pitch

The Chinese hackerspace Chaihuo (柴火) is not the kind of place in which one would expect to find the prime minister of China. It is a 15-by-10 square meter room on the second floor of a refurbished factory building located in OCT Loft, a creative industry park in the city of Shenzhen.[21] Chaihuo is a community space that provides its members with access to low-cost machines, electronic tools, open source hardware platforms, and educational kits for a small monthly fee (RMB 50[22]—at the time, less than USD 10). On the weekends, it frequently hosts educational workshops that are open to the broader public; at these workshops, attendees can learn how to reverse engineer the printed circuit board of an electronics toy or how to use the open source microcontroller platform Arduino to build their own DIY robots. Chaihuo is in many ways a typical hackerspace. On a regular day, one finds its tables covered in electronics tools and components, breadboards, alligator clips, and wires spilling out from work-in-progress prototypes, giving off a vibe of unhinged creativity and messy experimentation. Hackerspaces (or makerspaces as they are also often called) had rapidly proliferated across the world since the 2007/2008 financial crisis, and by 2015, Chaihuo was one of several thousand hackerspaces worldwide. These hackerspaces were not the spaces frequented by the type of "hackers" portrayed in movies—basement apartments littered with Mountain Dew cans where young men illegally accessed information or broke through security barriers. The hacking that took place in Chaihuo and other hackerspaces like it was a project of self-transformation from a passive consumer into a "maker"—an active participant in social, economic, and political processes, made accessible via technological tinkering.

As unlikely as it seems, though, China's prime minister, Li Keqiang, did indeed visit Chaihuo on an official state visit in January 2015, alongside two other Shenzhen-based businesses: the tech giant Huawei and a renowned investment bank. "Makers," the prime minister declared during his visit, "show the vitality of entrepreneurship and innovation among the people, and such creativity will serve as a lasting engine of China's economic growth in the future."[23] Shortly after the visit, the Chinese state newspaper, *Xinhua News*, publicized the prime minister's praise of making, and of Chaihuo in particular; it ran an article with the headline 李克强鼓励"创客"小伙伴: 众人拾柴火焰高 (Li Keqiang guli "chuangke" xiao huoban: zhongren she chaihuo yangao),[24] which loosely translates as "Li Keqiang encourages young 'makers': everyone should ascend to excel like Chaihuo."[25] The Chinese government website posted a photo essay of the prime minister surrounded by young Chinese men showing off their latest technological creations; later, the government website asserted that this visit showed that the prime minister himself was a maker at heart, for as a politician, he (like a maker) was approachable and invested in empowering citizens to "make things happen": 创客李克强: 创造一个让人时时感到方便的政府 (Chuangke Li Keqiang: chuangzao yi ge rang ren shishi gandao fangbian de zhengfu).[26]

Only a couple of months after the prime minister had returned to Beijing, he announced a new national policy that escalated making into a nationwide project. The policy was written around three key terms; "mass makerspace

FIGURE 1.1. Photos posted on the official Chinese government website (English.gov.cn/premier /photos) depicting Prime Minister Li Keqiang's visit at Chaihuo makerspace in January 2015.

众创空间 (zhongchuang kongjian)," "mass entrepreneurship 大众创业 (dazhong chuangye)," and "mass innovation 万众创新 (wanzhong chuangxin)."[27] These terms framed the policy's key directive, which called upon provincial, municipal, and district level governments to allocate existing resources toward setting up "mass makerspaces" modeled on Chaihuo. The policy was aimed at developing these makerspaces for "the people" in order to help cultivate a "maker spirit" among the masses, a spirit that would in turn help foster an attitude of "self-making" and "self-entrepreneurship."[28] In other words, the policy held up Chaihuo as a model for the nation. About a year later, China allegedly housed the largest number of makerspaces globally—several thousand such spaces were reported to have sprung up across the nation.

At first, the official endorsement of making by the CCP (Chinese Communist Party) might strike you as counterintuitive; why would the CCP, often associated with top-down decision making, authoritarian rule, and harsh limitation of personal freedom, endorse the workings of a small makerspace, associated with prototyping alternatives to entrenched structures of technological control and injustice? You might read the CCP's appropriation of making as yet another example of how the very tactics of grassroots intervention leveled against "the system" turned out to feed and sustain it;[29] or as demonstrative of how authoritarian rule allowed for the fast and efficient implementation of new policies, deemed impossible for Western liberal democracies. Perhaps you see it as a success story of a grassroots movement that managed to shape policy making and educational agendas on the national level.

While each of these interpretations gets something right, they each also occlude something; they assume making as a universal project—as standing broadly for what was, however, a very specific, American-centric articulation and manifestation of making that China's maker and open technology advocates felt pressure to take up as a reference and a model for their own work and ideals. When China's prime minister urged citizens to model themselves after Chaihuo makers, he endorsed a particular version of making—the kind that was marshaled by some of America's most well-known maker and open source hardware advocates and that the Chinese founders and members of Chaihuo had worked hard to be associated with in order to be "seen" and taken seriously as equal and creative partners by their Western counterparts in the maker movement (a theme I unpack at length in chapters 2 and 3). Indeed, it was this dominant articulation of making that centrally shaped its broad uptake across seemingly opposing and contradictory viewpoints (including the endorsements of making by the Chinese government)—a thread that I turn to next.

In 2015, the same year that China's prime minister had visited Chaihuo, the magazine *Pacific Standard* published an article by Dale Dougherty titled "The Future of Work: Join the Maker Movement." In this article, Dougherty, who had founded the San Francisco–based publication house Maker Media (which helped proliferate the ideals of a global maker movement), argued that by becoming "makers of machines," individuals could regain control over their lives and over work amidst the spread of automation and precarious economic conditions:

> Losing faith in the utility and even beauty of machines is losing faith in the kind of future we can build. We need more machines and even smarter robots that can do more for us. Even if such machines do eliminate jobs, they also create new opportunities—or at least new problems to solve. . . . If we see ourselves as the makers of machines, we are invested in creating the future, rather than having it imposed on us. The big challenge is how to get more people, not just a few, to take advantage of opportunities to do work that matters to them and makes the world a better place.[30]

Dougherty's article centered on the story of Lisa Fetterman, a passionate hobbyist cook who, fed up with the steep prices of sous vide cooking machines, took matters into her own hands and produced her own machine: she designed, prototyped, and manufactured it on a mass scale. Along the way, she transformed herself from a maker of food (often feminized labor) into an entrepreneur (often rendered masculine). By doing so, Dougherty argued, she also transformed work itself into something that was no longer drudgery, but desirable and even fun:

> Without any prior knowledge in hardware, she [Lisa] and her husband began building a sous vide prototype of their own . . . she moved to China for three months to learn how to do that. That's how her company, Nomiku, was born. Work, she told me, "was once boring but now I really don't feel like I am doing work."[31]

Dougherty's narrative condenses a key promise of the maker movement: self-transformation into an entrepreneurial agent was now democratized, available to everyone. Makers need not be professional engineers, designers, or computer scientists to build machines and devices, start tech companies, sell products, and shape industries along with the future of work. Makers, instead, could decide for themselves what their technologies looked like. They would no longer have to listen to the state, the corporation, or the university to tell them what the good jobs were, what the good life meant, and the kinds of futures they wanted. Instead, they could prototype their own. Making offered both a sense of control over work—a way to escape what Dougherty saw as the drudgery and feminization of labor—and, more important, a way to intervene in the structures of neoliberal capitalism typically seen as inevitable.[32] Dougherty's article was published as debates about the ethics and moral dilemmas of the tech industry were escalating across American news and social media. In 2015, AI, big data, and the Internet of Things were becoming fodder for rhetoric of both promise and fear. Standards like 5G and machine learning approaches in AI were portrayed as both inevitable and necessary, as delivering on several decades of promises about the smart self, the smart city, autonomous transportation, and interconnected living. Yet it was feared that these developments would have drastic, poorly understood impacts on work, from deskilling to outsourcing. They seemed to put at risk jobs that thus far had been protected from labor changes—white-collar jobs such as tech work and creative work.[33]

Following the global financial crisis in 2007–2008, such forms of crisis thinking have become pervasive in the United States in particular;[34] indeed, our contemporary moment is often articulated as one of permanent crisis characterized by ever-increasing economic uncertainty, political instability, and severe environmental havoc. But projects of "economic development" and modernization (and the resource extraction, displacements of humans and animals, and extinction they have legitimized and demanded) of course had already much earlier outgrown their laboratory Third World.[35] The financial crisis made visible the processes of exploitation, the feminization of work,[36] and the neoliberal colonization and economization[37] of ever-larger swathes of our lives,[38] for they began to affect people who had been temporarily (even if precariously so) insulated from them. Workers in tech industries, design consultancies, architecture firms, and computing research labs began to question the promise of modern progress that had sustained their own work.[39] The years following the crisis were marked by a reckoning with the complicity of technology, engineering, and design in labor exploitation, resource extraction, economization, and the creation of what critical race scholar Neferti Tadiar calls "surplused populations"—"populations figured as forms of bare life, at-risk populations, warehoused, disposable people, urban excess,"[40] who serve capital as it moves from one site to another. American news media coverage began warning about the costs of ruthless neoliberal capitalism and the precarious work conditions it created, with pieces such as "Entrepreneurs Are the New Labor" in *Forbes* in 2012,[41] "The Future of Work and Workers" series run by *Pacific Standard*,[42] *Fortune*'s 2016 "Even the IMF Admits Neoliberalism Has Failed"[43] by Ben Geier, and the 2017 *New Yorker* article "The Gig Economy Celebrates Working Yourself to Death" by Jia Tolentino.[44] Increasingly, people criticized Silicon Valley as a central culprit in this destabilization. *Wired* magazine, known for its central role in propagating and legitimizing the belief in techno-optimism in the 1990s,[45] began running articles on the pervasive sexism of the Silicon Valley industry as well as essays calling out the tech industry's complicity in neoliberal capitalism. These writings signaled the dilemma of a tech industry that found itself confronted with a growing public suspicion that its countercultural ideal of "tearing down hierarchies, undermining the sorts of corporations and governments that had spawned them, and, in the hierarchies' place, create a peer-to-peer, collaborative society, interlinked by invisible currents of energy and information,"[46] had benefited only a small minority.

The idea of a "maker movement"—while couched in a rhetoric of global and universal applicability—came out of this specific moment of techno-crisis thinking that was particularly pronounced in the United States. The idea that if individuals turned themselves into makers of machines, devices, and tools, they could also prototype concrete alternatives to contemporary capitalism and the spread of loss of control, vulnerability, and insecurity, was specific to the American context following the global financial crisis, accompanied by a rising distrust—expressed by Americans themselves—in their own IT industries. While ideals of open source hardware and electronics hacking had been circulating since the mid- to late 1990s, they picked up steam in the years of the financial crisis, 2007–2008. In 2008, the flagship Maker Faire in the Bay Area, an annual gathering of maker and open source hardware enthusiasts, began counting participant numbers over 200,000; during the same year, the number of hackerspaces in existence worldwide rose from less than a hundred to more than a thousand. Many of the key ideals of hacking—the notion that societal structures are technological and as such modifiable via technological tinkering, the idea that tinkering with code could be a form of moral and political intervention[47]—were rearticulated in terms of "making" and thus rendered as available not only to geeks or computer scientists but to everyone. The global spread of hackerspaces and the growth of Maker Faires proliferated the sense of being part of something greater.

It was the writings, educational initiatives, and methodological toolkits promoted by a group of powerful, mostly male actors[48] associated with networks of tech production and innovation in the United States that attached this affect of intervention to making. They "pitched" making as a democratized approach to technology production and design. This democratization of tech production, they argued, would enable "everyone" (rather than just the elite) to turn themselves into self-defining and entrepreneurial agents of change who intervened in societal and economic structures not by political action and mobilization, but by technological experimentation. In this vision, makers "take action" by experimenting and tinkering with technological alternatives; they *prototype* alternatives to the status quo rather than demanding change by political activism, critique, social uprising, or protest. In this pitch, they appealed at times explicitly and at other times tacitly to socialist values and ideals. As a telling example, we can turn, for instance, to a quote from the book *Makers: The Next Industrial Revolution* (2012) by Chris Anderson, former editor-in-chief of *Wired* magazine:

> the ability—to manufacture "local or global" at will—is a huge advantage. That simple menu option [of a digitized machine like the 3D printer] compresses three centuries of industrial revolution into a single mouse click. If Karl Marx were here today, his jaw would be on the floor. Talk about "controlling the tools of production": you (you!) can now set factories into motion with a mouse click.[49]

Anderson's invocation of Karl Marx here exemplifies portrayals of making through images and ideals of political action and mobilization. Not without irony, socialist ideals are reduced to serve the function of a pitch, i.e., the formulation of an idea or project as attractive to potential investors and media by producing an affect of anticipation and by promising intervention at scale. Anderson's book, like other prominent writings on making from the periodical *Make: magazine* to Neil Gershenfeld's *Fab* and Mark Hatch's *The Maker Movement Manifesto*, all "branded" their arguments by invoking socialist imagery, tactics, and language. They claimed that making enabled societal and economic change for "everyone." They called making a movement, they wrote maker manifestos and maker bills of rights. They articulated making as returning control to "the people" and as democratizing peer production, open sharing, and co-ownership of resources and knowledge. They talked about acquiring "ownership" over the means of production, intervening in corporate control over economic processes, from commodification to finance speculation. Unlike Marx, though, who understood alternatives to capitalism as emerging from class struggle, solidarity movements, and the collectivization of workers, these writings portrayed *individual* self-actualization as desirable. Notably, Anderson describes making's transformative power as stemming not from a collective "we" but from many individual "you's."

These writings and projects construed one of the core techniques of neoliberal governance—the framing of life itself in economic terms—as desirable and as key to enabling social change. They produced what seems to be an inherent contradiction: a *socialist pitch.* "Pitching" is typically associated with a start-up's ability to formulate itself as attractive to venture capital, often using a standardized script, for the start-up has to produce a feeling of anticipation and a promise of scale.[50] The "socialist pitch" derives its power from this technique of producing an affect of exuberant excitement and buy-in, by drawing specifically on the language and image of change and justice. Pitching works through promise rather than actual change; investors fund ideas and prototypes of intervention rather than finished products. Promises are "expressions of desire," feminist and critical race scholar Sara Ahmed reminds us. They "point us somewhere, which is the where from which we expect so much . . . for something to be promising is an indication of something favorable to come." The socialist pitch is aimed not at social justice per se, but at creating a desire for change and a feeling that justice via technological intervention is the only path forward. By the socialist pitch, I do not mean to suggest that the American advocates of the maker movement made a case for building a socialist or communist society. On the contrary, socialist imagery, tactics, and ideals were utilized to make a rhetorical move, to recuperate technological promise and modernist ideals of progress, precisely as people realized that the good life and the modernist dream of technological progress and solutionism had, in fact, long been unattainable for most. Socialist ideals were used to attach an affect of intervention—via making—to self-economization, i.e., the neoliberal demand that one convert the self into human capital, investing in various aspects of one's own life in order to make the self attractive to the machineries of finance speculation and investment.[51]

Processes of economization, i.e., the framing of humans, animals, the environment, and life itself as economic,[52] are at times simply taken as the consequences of neoliberalism. One of the great myths of the neoliberal ideology is that market capitalism is laissez-faire, that economization of life simply happens with no intervention. I use the concept of the *socialist pitch* to show that the economization of life and the creation of human capital had to be actively cultivated.[53] I focus on the artifacts, instruments, machines, people, and sites that imbued processes of economization with affect.[54] In chapter 4, I show the role that incubators and adjacent entrepreneurial training programs (from startup weekends to hackathons and accelerators) play in training people to see themselves as human capital and to channel their commitments to justice and technological alternatives as attractive to finance capital.

The socialist pitch remoralizes economization (of the self, life, the environment, and things) by rendering it as key to an optimistic, interventionist, and future-oriented way of living. The economization of life is portrayed as desirable and as providing individuals with interventionist capacities; *economic life* appears to be *entrepreneurial life*. The socialist pitch thus functions as a *market device of finance capital*. Market devices are typically defined as the technological, discursive, and/or human actors that generate knowledge and practices that create markets and thereby define their means of commercial exchange.[55] Pricing techniques, accounting methods, monitoring instruments, trading protocols, benchmarking procedures, and economists all have been shown to function as such devices to create markets.[56] These market devices operate by translating complex societal and political processes via quantitative measures and simplification into manageable and seemingly controllable entities.[57] By contrast, market devices of finance capitalism operate through affect and by channeling yearnings and aspirations. They imbue processes of economization with feelings of actionability, intervention, and the promise of happiness.

The broad endorsement of making further legitimized the technique of the socialist pitch that remains pervasive in the tech and creative industries despite a growing suspicion of these industries' promises of better futures and the good life; when users participate in digital platforms such as Amazon, Facebook, or Uber, they are celebrated as entrepreneurial agents of content creation, remix, and even social movements, masking their transformation into co-creators of economic value behind a story of empowerment;[58] when citizens are celebrated as entrepreneurial change agents, the demand placed on them to construct markets and innovate national economic development appears hopeful;[59] when educational reform is framed as an entrepreneurial endeavor, the broken promises of the techno-fix are re-invested with renewed optimism and feelings of social change;[60] and when tech companies propagate the mantra of disruption, acceleration, and breakage, the "people and places broken in the process" are enrolled in an enticing story of market development and progress.[61] All of these are processes of economization that reproduce and often intensify inequities and violence because they work behind the "socialist pitch" of participation, inclusion, and empowerment; they are also a process of depoliticization,[62] for the subjects interpellated through such participation are positioned in ways that discourage collective agency and resistance.[63] People's hopes, dreams, and yearnings for alternatives to regimes of exploitation and disempowerment paradoxically end up further enabling them.

The socialist pitch deployed by prominent maker advocates recuperated certain aspects of earlier revolutionary rhetoric common to innovation discourse.[64] It can be understood as a rhetoric of techno-optimism haunted by a shift in attitude toward digital technology—increasingly cautious, reflexive, and ambivalent. It was this socialist pitch that engulfed making in a feeling of possibility, an imaginary of action-ability that circulated through various and often even opposing sites and places. And it was this "socialist pitch," the promise of democratized interventionist capacity, that masked the universalizing tendencies of the American maker discourse and practice. Articulated as an extension of earlier Western traditions from the American Internet counterculture to European ideals and practices of craftsmanship to the democratic ideals of post–World War II design—from the experiments at the Bauhaus to Black Mountain College,[65] making was posited as having arrived "here" (in the West) first, while then proliferating outward and made specific in local contexts. While making was positioned as intervention into persistent structures of technological elitism and control, the writings on making and open source hardware that proliferated in and beyond the United States largely retained a very old ideal; the West as the emanating center of future making,[66] following in the footsteps of a long Western lineage of scientific and technological experimentation, hacking, design, and craftsmanship.[67] What endured, in other words, was an old, and all-too-familiar colonizing narrative, a "universalizing view that promotes a notion of technological and scientific progress" that in its claims to universality masked how it was deeply entwined with specific national, state, and commercial interests.[68]

Let me pause here and return to the region that has come to hold a paradoxical place in this old, new project of future making—Shenzhen. Remember Lisa Fetterman, the hobby cook who made her own sous vide machine? There was an element of Fetterman's story that one could very easily overlook, and that was, indeed, mentioned only in passing by Dougherty himself. Lisa Fetterman "moved to China" to accomplish her self-transformation from passionate hobby cook to savvy tech entrepreneur and maker of machines. But why would an American entrepreneur have to travel to China to free herself from the drudgery of work? Why did the promise to prototype the future of work in America hinge on China and, more specifically, on Shenzhen, where Fetterman traveled?

Shenzhen has long figured in the Western imagination through the sensational news stories of Foxconn worker suicides and copycat electronics production. It has also long been considered Silicon Valley's unimaginative counterpart, as the site of mere execution of ideas "created" elsewhere, a place that was backward and lagging behind the "forward-looking" centers of tech innovation. While this image of Shenzhen persists, another narrative of the region emerged around 2010 reformulating the region's "backwardness" as an opportunity (rather than what held the region back). This happened for two reasons: first, China's early open source and maker advocates had begun to form as a loose collective between several Chinese cities in 2007, and they had turned toward making as a way to reposition China in the global imagination, shifting it from being seen as a low-quality producer to being an equal partner in hardware innovation (I explore this in detail in chapters 2 and 3). Second, foreign designers, engineers, artists, educators, and entrepreneurs began travel to China (mostly Shenzhen), with numbers peaking in the years of 2013–2016. Among the crowds who came to China were well-known figures of the American maker and open source hardware scene as well as investors, entrepreneurs, designers, scholars, artists, and educators from elite institutions in Silicon Valley, the East Coast of the United States, and various corners of

Europe. Their stories of working with Chinese factories, China's makerspaces, and open source hardware advocates were documented on personal blogs and eventually by an expanding number of Western news media outlets (from *Wired* magazine to *Forbes* and the *Economist*).

Together, these accounts produced an image of Shenzhen as a rising innovation hub, a "Silicon Valley of Hardware," and "Hollywood for Makers." It was in this moment that Shenzhen was enrolled in the socialist pitch advanced by America's maker advocates, investors, and corporate players by attaching an affect of intervention to the city and its wider region. The "travel reports" from (at first largely Western) designers, educators, engineers, artists, maker advocates, and even scholars portrayed Shenzhen as a new frontier and the next "regional advantage"[69] not unlike what the West Coast had once represented for the early Internet counterculture, a "place where things still get made," as Dougherty himself put it when I interviewed him in 2014 during his first visit to the region. Not uncommon across such accounts were colonial tropes of adventure, frontierism, and of "going back in time." Within the time span of only a couple of years the story of a "new China" was constructed—China as a prototype nation that was promising for the entrepreneurial designer and engineer, exactly because it was construed as "other" than the West, i.e., because it was seen as a site of fakes, copies, violations of IP regulations and copyright law, and lax rules of law and regulations writ large (see chapter 3 for details). Shenzhen was portrayed as an opportunity to go back in time and as the "underbelly" of the glittery world of Silicon Valley tech innovation, i.e., a place where the "cultural hegemony"[70] of the global Intellectual Property Regime (IPR) and the black-boxing of technology were not (yet) fully accomplished and complete. It was celebrated as a place where one could "see" the inner workings of industrial production: large-scale machineries and the "hands-on" labor on the assembly line were celebrated as providing opportunities for the designer and engineer to move beneath the slick surface of the software interface and re-learn the craft of production. It was framed as Silicon Valley's "other," a new frontier, and a "scaled-up" version of the hackerspace, i.e., a "city-size" laboratory to deliver on the key promises of the maker movement as I described it earlier: to prototype alternatives. Embedding oneself into Shenzhen, so the vision, would provide technology producers, researchers, designers, and activists with the tools necessary to intervene across scales by moving outside the "clean" and "elitist" office spaces of "venture labor"[71] and creative work. Shenzhen was understood as a place where one could travel and "see" scale in action; global supply chains, mass production, the city as special economic zone, international borders, and global ports of trade. One could "see" and thus understand acceleration and economies of scale—one would learn and thus be empowered to intervene in the workings of capitalism.

FIGURE 1.2. Still image—*Wired* UK documentary, 2016: "Shenzhen: The Silicon Valley of Hardware." From *Wired UK*.

Frontiers, anthropologist Anna Tsing reminds us, "are not just discovered at the edge; they are projects in making geographic and temporal experience."[72] Frontiers are places where one goes to see and build the future, and to erase certain pasts.[73] China was "(re)made" in the broader tech imagination as a place to dream, to see the future "again," precisely as that future was being called into question. It was rearticulated as it dawned on many people active in the worlds of (largely Western-centric) technology innovation, research, and design that the dream of modernity to "bring about an end of scarcity, an abundance of goods, permanent employment, prosperity and the fulfillment of personal happiness"[74]—in other words, the dream of "living the good life"[75]—was exclusionary from its inception; it had not been, and would not ever be, attainable for most. It was in this moment that the south of China, and the city of Shenzhen, in particular, came to be seen by many as a paradoxical "laboratory" of exuberant scale, where the master could dwell in the illusion of taming the land just a little bit longer. In Shenzhen, one could relive—even if only temporarily—the modernist promise of progress and control that had once made Europe and America great.

It is this displacement of modern, technological promise onto Shenzhen (via the socialist pitch of entrepreneurial living) that co-produced China as a prototype nation and that explains the Chinese government's absorption of the ideals of a tiny makerspace in Shenzhen into a nationwide policy and educational initiative—the part of the story with which I began and return to in greater detail in what follows.

Prototype Nation: Histories of the Future

> Images of the past help facilitate a vision of a future that harkens back to
> aspirations from the past.
>
> —JUNO SALAZAR PARREÑAS, *DECOLONIZING EXTINCTION*, P. 58

The displacement of technological promise I have described so far is not a linear move from "here" to "there" (from the West to China) but works through
temporal and spatial contingencies. Specifically, the making of Shenzhen as a
new "frontier" was co-produced on the one hand by anxieties about the short-
comings of modernist ideals of technological progress (anxieties that registered
in the promise of making and a growing suspicion of the tech industry particularly pronounced in the United States and Europe), and on the other hand by
long held desires of Chinese people to overcome racialized othering, shaped
by the CCP's own ambitions to integrate citizens into "the dream of regaining
China's stature as an empire" and to "attain material and moral parity with
the West,"[76] a project that has occupied Chinese leaders since China's partial
colonization by Europe in the nineteenth century. In other words, colonial
formations of the past govern both of these political projects; the mobilization
of insecurity, fear of loss, and crisis in the West, and the channeling of desires
for parity and national sovereignty in China. Colonial pasts reverberate both
in the contemporary displacement of technological promise (e.g., Shenzhen
portrayed as Silicon Valley's "other" as I described it earlier) *and* in political
ambitions to reposition China as a forward-looking, happy, and optimistic
nation that reasserts itself globally—a project that has gained force since Xi
Jinping became president in 2013.

This book shows that making was appropriated by the CCP to mobilize
feelings of optimism and happiness on a mass scale in the very moment that
the party feared social and political instability due to China's first significant
economic slowdown since the economic opening reforms in the 1980s. The
party feared that the slowing economy—what it has referred to as "China's New
Normal"—would lead to social instability. The CCP had retained its legitimacy
in part due to its assertions that it had lifted millions of people out of poverty and it feared that people's dissatisfaction with the economy could harm
the legitimacy of the party state.[77] The particular version of making that the
prime minister endorsed when he visited Chaihuo makerspace in Shenzhen
in 2015 was ideal for the CCP in this particular moment; the socialist pitch
(as advanced by America's maker advocates and as I outlined it earlier) had
translated economic life into entrepreneurial life, i.e., it had framed processes
of self-economization as hopeful and as democratizing technological agency.
For the party, this pitch constituted an ideal *technopolitical instrument of affect*,
i.e., a tool to frame its own political agendas via a language of technological

promise.[78] Specifically, it helped position the political demand placed on individuals to self-economize on behalf of the nation (to address "China's New Normal") as advancing long-held yearnings of Chinese citizens to be seen as modern innovators. Making, in other words, was ideal for the CCP to portray what was fundamentally a neoliberal strategy—the demand of citizens to self-upgrade into optimistic, economic agents who drove innovation and who built their own jobs rather than relying on the state—as in line with the principles of the communist party and as serving the hopes and dreams of the people.

Just a couple of months after the official endorsement of making by China's prime minister in 2015, I attended the Pujiang Innovation Forum in Shanghai, where I listened to a keynote by Wan Gang, the then-minister of science and technology, in which he encapsulated the CCP's utilization of making: "In China's New Normal, makers, open source and open hardware—as a form of entrepreneurship amongst the people—can help realize China's innovation strategy. It is the opportunity of the majority, rather than just the privilege of the few, to realize a lifelong dream." Not unlike the socialist pitch of America's maker advocates, the Chinese minister of science and technology here deploys socialist language (the majority, the people) to mobilize people to take individual action. Further, the language of entrepreneurial agency and innovation is strategically paired with one of the key (national and global) branding strategies the CCP has deployed under Xi Jinping: the Chinese Dream.[79]

Since he ascended to power, Xi has positioned China as a nation of dreamers, a place of promise and happiness. This includes not only Xi's notion of the "Chinese Dream," but also a series of "happiness campaigns,"[80] and his appropriation of the citizen-driven phrase "positive energy" (zhengnengliang 正能量).[81] These constructs indicate a discursive "shift from locating the future outside China (by figuring China as backward and the West as advanced) to see[ing] China itself as the future."[82] They are aimed at creating an affective relationship of mutual interest and "positive feeling" between the Chinese party state and citizens. This affective bond is aimed at advancing the nation and at solidifying the party state as *the* political power to support the Chinese nation and its people. The happiness campaigns, anthropologist Jie Yang shows, "encourage people to focus on the self, adjusting oneself to realize one's self" on behalf of the nation. The pursuit of happiness, in other words, becomes a moral imperative for the "quality" citizen who advances the self to advance the economic future of the nation; as Jie Yang puts it, "how better to legitimate crippling economic restructuring and intensified social stratification than to deploy programs that suggest that these processes are actually an opening that could lead to happiness?"[83]

Making proved ideal for the CCP to portray this neoliberal ideology of self-care, self-realization, and self-enterprise (in the name of happiness) as advancing people's own yearnings to be seen as innovative, creative, and modern—a subjectivity that had long been denied to most of China's citizens, both by their own government and by the West. As I explain in greater detail in chapter 2, the CCP has strategically utilized China's partial colonization by the UK and France during the Opium Wars in the nineteenth century—what the party refers to until today as China's "period of national humiliation"—to invoke in people desires to upgrade into modern, "civilized," and "quality" citizens (and to legitimize a range of governance techniques, including the adoption of "neoliberal exceptions"[84]). The discourse of China's citizen's supposed lack of civility emerged from China's partial colonization and has been variously deployed by China's leaders ever since to argue that "China's humiliating and unequal participation in the globalized historical time of the modern instate system"[85] (its failure to modernize) was due to the failure of its people to modernize, owing to people's lack of "civility." The cultivation of a "civilized" citizen (a citizen recognized both as uniquely Chinese and as modern by the West) has been at the heart of China's modern nation-building projects ever since, from the late 1920s until today. China's projects of modernization have always struggled with the fact that the very notion of modernity was itself a colonial imposition.[86] When the CCP came into power, it argued that socialism constituted an ideal approach to establish an alternative modernity to the capitalist West.[87] "The socialist revolution itself," Dirlik and Meisner explain, "was the product of the deepest urges of a society to gain entry into the stream of history as its subject against forces that denied to it such entry."[88] The cultural revolution did not reject development; it sought to restructure the idea of development by politicizing it.[89] Modernity, in other words, was to be achieved not through economic expansion but by cultivating revolutionary subjects via social struggle and mass mobilization.[90]

Following Mao Zedong's death, the CCP positioned China's economic opening reforms in the 1980s as necessary to address China's failure to modernize during the previous decades. Political leaders at that time (Deng Xiaoping being the most prominent) argued that this was to be accomplished by transforming revolutionary subjects into citizens of "quality" (素质 suzhi), invested in their own economic development for the purposes of furthering the nation. The notion of quality (suzhi) was deployed to attribute China's failure to modernize under Mao Zedong—once again—to the "low quality" of its people. The economic reform period was characterized by a turn away from politicization to economization. Modernity was no longer portrayed as arising from ideology but a pragmatic, fact-based approach—an attempt "to seek truth from facts" and a "socialism with Chinese characteristics" as Deng Xiaoping put it in the 1980s. The reform era was a "pragmatic adjustment of revolution," i.e., socialist progress and social change had to "follow the demands of economic development."[91] The party portrayed self-transformation of citizens into "economic subjects" as key to this adjustment. In other words, the political aspiration to return China to its "rightful place" and guarantee sovereignty from Western hegemony that had guided postcolonial governance was now to be accomplished via what during the cultural revolution had been punishable by death: self-investment, entrepreneurial activity, privatization of state-owned land and resources.

How did the CCP get people to perform this drastic self-transformation from revolutionary subjects into economic agents? How did it convince people to give up socialist support structures such as the "iron rice bowl" (the guarantee of life-long employment) and the danwei system (the "work unit," i.e., work-based communes that provided living space, meals, medical care, socialization, and ideological indoctrination, all in one small geographic area)? To "reform" its people, the CCP established so-called Special Economic Zones (SEZ), i.e., spatially bounded zones in which economization, privatization, and foreign direct investment were encouraged, while they remained at first still prohibited (or at least not enabled) in the rest of the country. In 1979, Deng Xiaoping declared Shenzhen a SEZ (alongside Zhuhai, Shantou, and Xiamen)—a laboratory to "feel out" how far away from socialist structures and values China could move without changing its essential character.[92] Crucially, this "laboratory" model induced desires in citizens for economic and social upgrading. Political scientist Mary Gallagher, for instance, shows that the 1980s' "dual track" system of the special economic zones that allowed for socialist models and organizations to coexist alongside the new experiments with capitalist markets created competition over FDI (foreign direct investment) between regions and cities.[93] It led to a race to implement more flexible labor policies, to create a mobile workforce, and to grant autonomy to enterprises. By "allowing some to get rich faster," as Deng Xiaoping had famously put it, the government induced desires to self-transform and embrace values such as autonomy, self-reliance, and economic initiative.[94] Economic reform was pushed ahead, while resistance was reduced—a "contagious capitalism."[95]

During the 1990s and 2000s, the CCP stimulated economic growth by expanding the technique of the SEZ rapidly throughout the rest of China (a well-known example is the SEZ of the Pudong New Area and the Lujiazui Finance and Trade Zone established in Shanghai in 1993). When China joined the WTO (World Trade Organization) in 2001, municipal- and provincial-level governments competed over receiving designations such as "creative city" or "city of design," which in turn would funnel resources from the central government in Beijing into their districts and provinces. Shanghai and Shenzhen, for instance, were among the first to turn old city neighborhoods into creative industry clusters and build high-tech innovation parks. These parks, zones, and refashioned neighborhoods (and the high-tech businesses and educational initiatives they attracted) were aimed, broadly construed, at cultivating citizens as "creative talent" (rencai).[96] The build-up of China's creative industry was motivated by the political ambition to mass produce "prototypic liberal subjects,"[97] people trained to model the social transformation desired by the CCP and "rebrand" the nation in both the national and global imagination as a creative producer.[98]

The official uptake of making by the CCP in 2015 has to be understood as prefigured by these various political projects that strategically invoke China's "humiliating" history of colonization, aimed at inducing desires in Chinese citizens to self-transform and self-upgrade on behalf of the nation. China as a prototype nation, as a nation that is newly emboldened and asserts itself as an alternative model of future making, is co-produced by these long-held aspirations to achieve parity with the West and the party's claims that to achieve a sovereign, modern Chinese nation necessitates a particular kind of citizen. When the prime minister of China visited the makerspace Chaihuo in 2015, the young Chinese men he met (and whom he framed as model makers for the nation) had already received international recognition and were regarded as legitimately creative in the Western tech scene. In the official state media's news coverage that followed his visit, the prime minister was depicted side-by-side Eric Pan and Kevin Lau, both active members of China's maker scene since the beginning. Pan had co-founded Seeed Studio, an open source hardware company that had been key to reformulating Shenzhen's image from a site of low-quality and copycat production into a legitimate partner in open source hardware and tech innovation. In 2010, Pan and Lau had founded Chaihuo—the organization that was key to hosting China's early featured Maker Faires (2014–2015), which drew hundreds of thousands of people, many from abroad. Both Seeed Studio and Chaihuo had become well-known entities in the international maker and open source hardware scene, celebrated by many prominent American maker advocates as advancing their ideals of playful experimentation and grassroots innovation in China, despite early Western accusations that China's version of open source hardware was copying them (see details in chapters 2 and 3). These Chinese men already had, in other words, transformed themselves into the kind of globally recognized, techno-optimistic, happy citizen subjects that the CCP is aiming to cultivate.

When China's prime minister endorsed Chaihuo, the aim was to induce desires in other Chinese to self-transform in the image of the model makers the prime minister celebrated—or, as state media had put it: "Li Keqiang encourages young 'makers': everyone should ascend to excel like Chaihuo." Modeling yourself after Chaihuo, in other words, promised Chinese people they would redeem themselves as creative producers on an international stage; if young Chinese managed to "excel" like Chaihuo, they would receive Western recognition and would be granted (by both the West and the Chinese government) the status of modern and happy world-class Chinese citizens. By transforming themselves into model makers, they would prototype at scale: lift up the nation and its image on both a national and global stage. This call for self-upgrade on a mass scale was aimed specifically at the marginalized and displaced in China's

FIGURE 1.3. Venue of the "2016 National Innovation and Entrepreneurship Week" (2016 年全国
大众创业万众创新活动周), also often promoted simply as "Maker Week" (创客周) in Shenzhen.
Banners at the venue and throughout the city promoted (in both English and Chinese) Shenzhen as "City of Makers" (与深圳同创造) and called upon citizens to "promote the development
of New Economy & Cultivate New Growth Dynamics" (发展新经济，培育新动能), to "Make
Innovation and Entrepreneurship Sweep across China" (创新创业，创响中国), and conjure
an "Era of Innovation. For Dreams of Entrepreneurship" (创新时代，创业梦想). Large screens
showed videos of Prime Minister Li Keqiang among crowds of passionate makers at the event.
Photograph by the author.

younger generation, for employment rates among China's college students
were low and upgrades in the manufacturing industry (like those carried out
as a result of the "Made in China 2025" initiative) had begun to drastically shift
the conditions of employment for a generation of young migrant workers,[99]
whom the government called upon to return to the countryside and become
entrepreneurs, starting businesses—ideally high-tech businesses—of their
own. This cultivation of Chinese citizens as human capital was crucial at the
moment as the party's leadership was focused on creating a positive image of
China abroad to create buy-in for one of its major, transnational infrastructure
projects, the Belt and Road initiative (BRI).

Shortly after president Xi Jinping ascended to power, he initiated two
major infrastructure projects, in part aimed at addressing China's New Normal
through industrial upgrade: the BRI and the "Made in China 2025" initiative.
The "Made in China 2025" initiative funded (to the tune of 2 Bio Chinese
Renminbi) the upgrade of China's old industries into intelligent/smart manufacturing zones, and the BRI aimed at moving China's capacities in industrial

production, real estate, and infrastructure development (train, roads, cities) into other regions in Asia, Europe, and Africa through a global trade and infrastructure route. Important for both of these projects was the cultivation of "human capital"—people who saw themselves as an instrument in advancing both China's image and its material infrastructures nationally and abroad. Making was ideal for the government purposes to induce in people desires to self-upgrade and build what the CCP referred to as China's "indigenous innovation economy" that would cement China's leadership in global supply chain markets, geopolitics, and the tech industry.

The CCP's invocations of China's colonial past *and* the colonial tropes of othering that wove through the displacement of technological promise onto Shenzhen co-produced "slow violence."[100] They occluded the violence of proliferating precarious conditions of life by harnessing individual dreams of modern belonging and yearnings for alternative ways of being. The promise to be granted the label of creative, modern producer led many of the people I met during my research in China to tolerate orientalist discourse and racism deployed (predominantly) by various Western actors *and* the precarious life their government demanded of them. My point is this: Technology research and design—especially in light of the recent and growing interest in ethics and politics of computing and design amid rising concerns over big data and AI[101]—must reckon with the violence that displacements of technological promise occlude and thus legitimize; violence in the form of racism, sexism, classism, and exploitation masked behind the promise of democratized tech innovation.[102] Just as colonial tropes of the frontier and of "othering" endure in a range of well-known technology practices and sites from Silicon Valley's exceptionalism and universalizing discourse and methods[103] to technology research programs such as ICT4D (Information and Communication Technology for Development) and Ubicomp (Ubiquitous Computing),[104] so are the recent displacements of technological promise onto what was formerly dubbed the "tech periphery" (from "smart zones" and "opportunity zones" in rural America to endorsements of certain regions and cities in the Global South as authentic maker cities and emerging hubs of innovation) marked by what Ann Laura Stoler calls "duress," i.e., the "enduring fissures" and the durable marks of colonial pasts.[105] Communication scholar Fan Yang urges us to understand such processes of colonization not strictly in terms of occupation of territories and the displacement of sovereignty. China, she argues, is not exempt from the conditions of American "coloniality" simply because it was never colonized by the United States. "Coloniality," she shows, is "a cultural logic that continues to exert influence through "imperialism without colonies."[106] Imperial formations and coloniality endure in technology production and design methods, including those celebrated as enabling inclusive and diverse futures such as making and its associated values of open technology and peer production.

The appropriation of making by the CCP was aimed at instilling in people desires to advance the party's ambitions to reassert China as a prototype nation, one that modeled for the world an alternative to the West, a China-specific approach to modernization and economic development—an "indigenous innovation" economy. While some of the people I met in my research expressed suspicion of the government's infatuation with making and open technology, many argued that the party was indeed supporting one of the key goals of China's maker advocates: to reposition China and its people as an equal partner in global tech innovation networks. There was a growing sense that China might in fact be the frontier, the place where the future was being made. And the party's espousal of these values only increased the people's affective connection to these goals. Because making was associated with play, experimentation, and tinkering—qualities that many Chinese I interviewed over the years insisted were Western, and more specifically American—they did not see the CCP's appropriations of making as part of the state's tactics of hegemony, for it functioned not by coercion but by promising happiness via self-transformation.

Differential Yearnings: The Labor of Promise and Future Making

> Yearning is the word that best describes a common psychological state shared by many of us, cutting across boundaries of race, class, gender, and sexual practice. Specifically, in relation to the postmodernist deconstruction of "master" narratives, the yearning that wells in the hearts and minds of those whom such narratives have silenced is the longing for critical voice.
> —BELL HOOKS, *YEARNING*, P. 27

What sustained these displacements of technological promise—of frontiers, of happiness, of future making—that came with no guarantees, one that might be deferred or withdrawn without notice? Who were the people who enabled others to live renewed technological promise, and what were the histories and stories of the places where they lived? This book tells their stories alongside those of the people who formulated and implemented the promise of making; it focuses on the sites and bodies that labored to sustain others' lives of technological promise. It was their labor, precarious and often hidden through gendered and racialized exclusions, that provided the necessary conditions to sustain the promise of technological progress, techno-optimism, and future making.

During fieldwork in 2013, I lived for several months in a modern high-rise building in Shahe, one of the city districts in Shenzhen. My apartment was right above the subway stop for "Window of the World," a forty-eight-hectare theme park with a 1/3 scale reproduction of the Eiffel Tower, which can be seen from across the district. The theme park was built in 1994 as a "world culture primer for China's political elite," who were eager to position Shenzhen as a "civilizational front line in the nation's efforts to 'join tracks' with the rest of the world." The name of the theme park referred to the SEZ itself, which China's political leaders at the time had called a "Window of the World." This move framed the SEZ (in the words of Jonathan Bach) as a "spatial threshold," which "mediates between China's economic space and that of other countries" and "through which one can look both in and out."[107] The theme park was built two years after Deng's famous return to the city in 1992, when he dubbed Shenzhen as a success and held it up as a model for the nation. The rhetoric of the "Shenzhen miracle," propagated by state media at that time, "concealed the precariousness and liminality that characterized and continues to define migrant workers' conditions," the historian Eric Florence reminds us. The people who were made responsible for remodeling themselves to follow Shenzhen's success "were the workers who came from rural hinterlands."[108] They bore the responsibility and precarity that came with the new lifestyle of the SEZ—what then-president Jiang Zemin in the 1990s referred to as the "Shenzhen spirit." This phrase describing the early SEZ was deployed to encourage the cultivation of a new kind of ideal subject, "a person able to transform her/himself and the socialist world," who would (like the SEZ) model the transition from socialism to capitalist expansion and economic development—a person who would live by the neoliberal doctrine, "decide for oneself, strengthen oneself, be autonomous, compete, take risks, and face danger." The construction of this model worker was key to the party-state's ability to "adapt its system of signs and symbols of socialism to the conditions of global capitalism," and to position Shenzhen as the passageway, the window into the world of global capitalism. The language framing Shenzhen as a laboratory and a window, which "emphasized the state's agency in 'opening' (a window) and 'conducting experiments,'"[109] is cemented into the day-to-day urban structures of China's contemporary middle class in the Window of the World theme park, which is now a tourist attraction, a site of leisure and consumption. The theme park materializes a vision of the SEZ not as an incubator for a range of ideas but as an example of successful party policy that set China on the path to the future.[110]

While I lived by the Window of the World theme park, the park was less of an attraction than the glittery shopping mall adjacent to it. With its myriad of restaurants, its large grocery store selling European brands and expensive, carefully wrapped fruits and vegetables, its international fashion labels, its Starbucks and Apple store, the mall extends into contemporary China the SEZ's technique of instilling desires for personal upgrade, autonomy, and self-actualization. The mall is the place where China's upper middle class consumes and plays, where those with sufficient suzhi (quality) feel like modern citizens enjoying the pleasures of their high-tech, modernist city. The thirty-six-story apartment building in which I lived was only steps away from the mall. My building, which was also home to the Chinese middle class who shopped at the mall, was taller than the replica Eiffel Tower across the street. Looking out from my apartment on the twenty-third floor, I saw high-rise buildings like mine stretching as far as I could see, an ocean sea of glittering lights amidst grey concrete.

This Bladerunner-like image fuels the postmodern sci-fi fantasies of Shenzhen that are invoked in the foreign travel reports depicting Shenzhen as a new technological frontier. But if one reoriented the gaze, away from the seductive draw of the glittering high-rise scape, another life world came into view; my building was located on the edge of Baishizhou, one of Shenzhen's few remaining urban villages. While the theme park, the mall, and my apartment building all contributed to the imaginary of Shenzhen as an "ideal modernist city" with no history, "a clean state, a tabula rasa,"[111] the "architectural form" of the urban village materially encodes the city's "rural history," its stubborn past. As anthropologist Mary Ann O'Donnell has painstakingly documented for more than a decade, Shenzhen's urban villages like Baishizhou played a central role in the making of the high-tech modern city, "provid[ing] the physical infrastructure that has sustained the city's extensive grey economy, including piecework manufacturing, spas and massage parlors and cheap consumer goods."[112] These villages—and their informal economies, based on what O'Donnell, Wong, and Bach (2017) call "illicit experimentation"—were the bedrock of the city's boom, providing affordable housing for the low-income migrants that helped build Shenzhen's economy. In 2007, Shenzhen's urban villages were slated to undergo renewal. Powerful, partially state-owned real estate firms benefited from erasing old neighborhoods and rebuilding them as more lucrative, upscale structures, from condominiums to office spaces and malls—like my apartment building and the mall beside it. These "renewal" projects treat urban villages as cankers—the city's past, its rural history, bursting through the image of the clean, modern, upscale city. As O'Donnell argues, the developers deny the villages' urban status, thereby legitimizing their erasure—and that of a particular past. While the glittering mall and the high-rise apartment complex represent the modernist fantasy of "a rationally ordered society where nature and society fit into precise categories and interact productively according to an unerring logic,"[113] Baishizhou represents its opposite: the narrowly built houses, the tangled electric wires that span its alleys, the small manufacturing shops around the corner of a wet market, make up a rich urban sociality, but officials see it as mud on the hem of the controllable, clean, logically ordered modern city.

I spent much of my time in Shenzhen in 2013 in Baishizhou. The urban village offered a different pace, moments of pause that sustained the projects of acceleration that I had come to study. As an ethnographer, I had joined a start-up team that had been admitted to a Shenzhen-based hardware incubator,

FIGURE 1.4. Street in Baishizhou—urban village, Shenzhen, 2013. Photograph by the author.

funded by a venture capital firm with deep roots in Silicon Valley whose Irish offices were strategically located for tax purposes. This three-month intensive training program invested in open source hardware and maker ideas that promised to scale, working with teams to bring their ideas to the prototype and pitching stage in an "accelerated" fashion (this notion of "acceleration" is an ethnographic thread that I explore in greater detail in chapters 4 and 5). This particular program invested in promising hardware products: data-driven smart systems, Internet of Things, smart objects, wearable technologies, and so on. In the evenings and on weekends, the start-up teams would often gather in urban village neighborhoods, for in the villages, people found a refuge from the demands to scale up and speed up. We ventured into these neighborhoods for fun and leisure, to eat at open-air bbq restaurants where locals played games and drank beer, sometimes staying late into the night. The villages not only offered space for play, they also provided access to affordable services, from mail delivery to electronics repair, laundry services, barber shops, wet markets, and clothing. Many of the team members of start-ups admitted to the incubator program lived on shoestring budgets, and the urban villages provided a crucial infrastructure to sustain the daily needs of team members under the precarious conditions of "venture labor"[114] at the incubator. In other words, the urban village's informal social, economic, and technological infrastructures (built, in part, by waves of migrants) enabled others to live out the promise of prototyping at scale.

In the offices of the incubator itself, a support network in many ways adjacent to that of the urban village made the work of the start-up teams and incubator staff possible—it kept afloat those entrepreneurial workers who were performing the desired work of technological promise, those who, under precarious conditions, worked to transform themselves and their economic and social positions into human capital on behalf of the investor. The incubator had hired two Chinese staff, women in their early twenties, whose key task was to help guide the start-up teams through the emotional ups-and-downs of their entrepreneurial labor in Shenzhen. As the European director of the incubator put it on the first day of the program, when "shit" was broken, it was their job to "fix" it and "to make our lives better." They were to stay in the background, to take care of the program's day-to-day functioning, and to ensure the start-ups' emotional well-being. As I elaborate in much greater detail in chapter 5, these two women were hired to perform what I call *happiness labor*—the work that goes into producing a feeling of optimism and cheerful delight, the affect that underpins entrepreneurial living. It is not unlike what Arlie Hochschild has described as "emotional labor," but it is more specific: it is the particular kind of labor relied on by a range of new organizational models in tech production, including but not limited to incubators, coworking spaces, makerspaces, and open innovation labs. This happiness labor is crucial to making bearable (and thereby sustaining) the precarious conditions of "venture labor";[115] this labor allows the entrepreneur to feel positive about technology despite his realization that for many, the self-investment on which he is gambling will not pay off.[116]

These differing sites and temporal scales—from contemporary happiness labor in pristine tech programs to the histories of urban villages and migrant labor that prefigure it—produce differential yearnings for alternatives, yearnings that feed the machineries of finance speculation and the displacement of technological promise. Each of these sites variously creates the foundations and possibilities for dreaming and prototyping at scale. The incubator, like many similar programs I encountered over my years of research, functioned like an educational boot camp. Its key aim was to train people to "pitch" or "dramatize" their dreams "in order to attract the capital they need to operate and expand." Anna Tsing puts it this way: "In speculative enterprises, profit must be imagined before it can be extracted, the possibility of economic performance must be conjured like a spirit to draw an audience of potential investors. The more spectacular the conjuring, the more possible an investment frenzy."[117] The start-up teams, the happiness workers, and the migrant workers who built Shenzhen, despite their drastically different positions in society, are all reduced through the logics of finance capital to their roles in the production of dreams. Some people dream; some people support those who dream. They have differential yearnings rooted in their respective positionalities and lives, but to investors they all look the same; they are relevant only if their dreams can scale; only the dreams that are dramatized or spectacularized are eligible for investment.

Some of these dreamers are held up as "models" to induce desires in others, and others are made invisible. The question of who is seen as offering technological promise and happiness and who gets to experience the affect of entrepreneurial living is highly gendered and racialized. During my time in the aforementioned incubator program, only one female entrepreneur was admitted, and only a few Asian men. As the start-up culture taught us, white boys pitch better. The (primarily white) start-up teams were future makers, people who lived the socialist pitch, people who were ascribed the capacity to make a difference, to intervene. They were charismatic figures, and many happiness workers and small entrepreneurs in the hinterlands dreamed of being in their shoes. Some yearnings are enrolled to enable other yearnings, some people's precarity sustains yet other precarious conditions of work.

It might be tempting to imagine the emergence of a unified, unmarked precariat that forms across these different subject positions. This book takes a difference approach. It brings together differential yearnings for alternatives, not to flatten them, but on the contrary to highlight how some people's dreams of social justice and change can actively weaken other people.[118] I show how the promise of individual empowerment and the consolidation of socialist ideals into the self leads to many individualized "you"s, who in competition with one another rarely find themselves in situations of collective solidarity. The mall by the Window of the World theme park, the middle-class apartment next to it, and the incubator together form an urban high-tech machinery of promise that's oiled for forward movement and for the making of the prototype nation; malls instill desires for middle-class life in the migrant worker, start-up incubators train people to see self-economization as providing them with hopeful agency, and makers model such self-investment for others rendered less innovative.

The concept of *happiness labor* contributes to scholarship of tech and digital labor by attending to labor exploitation in the sites celebrated as open and diverse, lauded for their commitments to peer production, participatory design, and open sharing. While recent scholarship has focused on how creative work has become highly precarious, there is much less attention to the multiple forms of precarity within the creative office, design studio, maker-space, or incubator. Labor exploitation is often assumed to occur elsewhere, on the factory floor, in the warehouse, in an Uber car; much less noticed are the workers who labor to maintain creative and innovative work itself, from cleaning staff to office assistants.[119] I spent much of my research with these workers, whose labor was never celebrated on stage or imbued with the kind of technological promise an entrepreneurial maker could claim, even though it enabled the work of creative production (see chapters 5 and 6).

I argue that attending to the labor that is necessary to nurture and support the neoliberal project of self-economization (of converting the self into human capital) challenges the notion that neoliberal processes and capitalist expansion are inevitable. This book shows that the subsumption of hope and of yearnings for alternatives by capital did not simply occur because of some inert law of the market, but had to be sustained, maintained, and nurtured. We must notice the labor that is necessary to sustain self-economization to refute the neoliberal ideology that economic growth and progress are inevitable and that resistance is unneccessary. If we refuse to participate in displacements of technological promise and attach ourselves instead to those bodies and sites that sustain them, we notice that technology can be otherwise; we recognize that the neoliberal doctrine of inevitability is already being challenged, on a daily basis, by those excluded from living its seemingly utopian promise.

A Few Words on Methods, Sites, and Position

The temporal and spatial contingencies I have sketched in this introduction weave throughout the book. With this approach, I align with the kind of critical historical analysis that moves beyond simply pointing to repetition (historical continuity) or rupture,[120] but highlights "partial inscriptions, modified displacements, and amplified recuperations."[121] And I align with the (largely feminist) anthropology of the global to attend to the contingencies of multiple sites and regions in uneven, contested, and negotiated alignments and frictions.[122] This is not a study of "tracing" the global proliferation of Silicon Valley's tech promises "outward" into the corners of its imagined periphery. On the contrary, this feminist ethnography of a global phenomenon unpacks how such modernist and colonial tropes of forward and outward movement, of promised connectivity, and transnational "flows" obscure the violence, exploitation, and vulnerabilities of the places and people who are made to endure the perpetual postponement of a more just and equitable world, despite and because of displacements of promise. The kind of multi-sited, genealogical,[123] and interscalar[124] investigation that I have outlined is vitally necessary, I argue: we cannot understand the broad appeal the ideals of the maker movement once had without also understanding the colonial pasts and presents that enabled it to rise, and we cannot understand China's shifting place in the world without accounting for how it was affected by the West's displacements of frontier thinking and technological promise.

By offering a translocal and multi-sited ethnography,[125] this book follows people, stories, and artifacts across regions; my research took me from China to the United States as well as to Taiwan, Singapore, Europe, and Africa. As an ethnographer, I followed the people who opened China's first makerspaces, co-working spaces, incubator spaces, and open source hardware businesses. I also worked with those who were hired to manage and sustain these spaces, but who were seldom interviewed by the media or given space on stage to "pitch" their work. Not all of the people I met were Chinese, and some of those from China had studied or worked abroad. Many who identified with making, hacking, or open source hardware considered themselves to be part of a global movement, traveling across regions and countries to attend events, make connections, present their visions, show off their creations, and learn from one another. In addition to long-term observations and interviews with a range of actors at maker/hackerspaces, coworking spaces, incubators, maker-related events, factories, and electronic markets, a central part of this work included the study of discourse, writings, and policy texts—not only those coming out of China, but also those circulating throughout what many of the people I met over the years described as a "global maker movement."

The research that informs this book spans a decade (2008 to 2018), from when I was a doctoral student, through my time as a postdoctoral fellow at Fudan University in Shanghai and the Intel Center for Social Computing in Irvine, and eventually to my transition to working as an assistant professor at the University of Michigan in 2014. Five years of this time, I lived and worked full-time in China. While my early work took place predominantly in Shanghai and Beijing, since late 2012, my research has frequently taken me to Shenzhen, when a growing number of open source hardware enthusiasts, makers, designers, researchers, investors, and start-ups just began traveling there. Shenzhen has transformed rapidly in the short time span, both in terms of its image and in terms of its material, economic, and political infrastructures. My research in Shenzhen in 2012 started out in a foreign-funded incubator and some of the Chinese makerspaces and open source hardware businesses that had just opened up. Part of this research included the participation in the day-to-day, decidedly unglamorous process of repeatedly pitching to investors and repeatedly revising the pitch until the central product idea was muffled in a cloak of tech buzzwords—the monotonous, vacuous language of techno-utopianism—that was required to get a VC's brief attention. I joined start-up teams as they visited local factories to source components, get prototypes made, or plan assembly production, and as they interacted with vendors on the Chinese ecommerce website Taobao to negotiate prices, shipping, and quantities.

Over the following years, I focused specifically on the transformations I saw unfolding at the intersection of urban and technological remodeling in Shenzhen. I followed the material and discursive productions that transformed the region in the broader tech imagination. This research included observations of and interviews with not only the growing number of Chinese start-ups, makerspaces, and incubator programs, but also with urban planners, policy makers, and government officials, and with the workers, engineers, managers, and designers in Shenzhen's manufacturing industry. As a white, female researcher, affiliated with an American university, conducting research on Shenzhen's manufacturing industry, I was often approached with a sort of bewildered curiosity that opened some doors and closed others. Not surprisingly, people at first associated me with the figure of the Western reporter focused on producing sensational sound bites, retelling the familiar story of China as a copycat nation or, more recently, telling stories about Shenzhen and China as rising innovation powerhouses. Many assumed at first that I would write negatively about Shenzhen's history of electronics production, and most people distanced themselves from the region's associations with copycat at first. But over the years, such interpretations of my work faded. Some began seeing me as an ally who could help correct China's image in the West—a project that, I came to discover, many were nevertheless highly ambivalent about.

For this period of research with Chinese workers, designers, and engineers in the manufacturing industry, I worked closely with Ingrid Fischer-Schreiber. She was an invaluable interlocutor for my thinking throughout my research and a key partner for these later parts of my fieldwork. Ingrid came to China for her training in Mandarin Chinese in the early 1980s and has since worked as a professional cultural interpretor, translator, and writer. Her long-term engagement with China and Chinese cultural and creative practices in particular greatly enriched my ethnographic research. Her outstanding language skills were of tremendous support during interviews and conversations with people from diverse ethnic and language backgrounds when my skills in Mandarin Chinese, obtained through training at universities and language programs in the United States and in China since 2006, would reach a limit. In all of these research engagements, my work has been centrally informed by the fields of science and technology studies (STS), cultural anthropology, human-computer interaction (HCI), and China Studies. I have read and cite widely across adjacent fields pertaining to my topic, such as political science, labor studies, postcolonial studies, critical political economy, communication and media studies, and southeast Asian studies. I am deeply committed to the critical sensibilities that inform these fields (and, ironically, suffuse much of the maker movement itself)—specifically, a modality of detangling, of unmasking concealed processes and ideologies, of challenging both linear time and the assumption that technology flows from a dominant Western "here" to another "out there."[126] I am both sympathetic to and wary of the well-known maker and hacker practices of reverse engineering and opening up the black box of technology. My commitments lie adjacent. I draw specifically from the techniques of feminist scholarship that have been working to untangle seemingly solid knots and show how technology can be otherwise.

My aim is to speak and write through the contingencies and complicities that held up the promise of making through a myriad of sites—financial investments, scholarly fascination, corporate interests, activist ideals, educational programs, techniques of governance. Specifically, I step away from two existing (dominant) interpretations of making—one of a successful grassroots movement that returns power to the users and consumers of technology, the other of how a once-authentic resistance movement has been co-opted by neoliberal corporations, states, and institutions. Both narratives depend on romanticized ideas about resistance and intervention, embodied in specifically racialized (white) and gendered (male) figures such as Edward Snowden or Steve Jobs. Stories of either success or failure prevent us from noticing what was most interesting about making; its capacity to accommodate diverse, often contradictory hopes and anxieties. This book does not speak of either pure dissent or total appropriation, but traces stories of ambivalent alliances and always already partially compromised ideals. We need to make sense of how the enduring seductive draw of technological promise can coexist with the proliferating distrust of its attainability. Future thinking and the promise of technological progress still exist in making, but only partially, and often contested. This book looks at efforts to work within and through neoliberal restructuring and technological promise, while moving outside their logics and attending to the contingencies and ruptures of techno-economic universals. Rather than seeking to pronounce making as ultimately "good" or "bad," as radical or thoroughly subsumed, I focus in this book on its tangle of contradictions.

In 2011, when I was a graduate student at UC Irvine, feminist technoscience scholar Donna Haraway came to speak as part of the university's famous Wellek Lectures series. In her talk, she invoked a figure that stuck with me, one that I have often returned to over the years: the cat's cradle, one of my favorite childhood games. In this game of "relentless contingency," the longer one plays, the more complex the patterns that are woven and pulled apart become, the deeper one becomes entangled. In order to function, the game demands continuous and often ambivalent attachment that is sustained through literal manual labor—the labor of one's hands.

This book traces such knotted attachments and complicit tangles, in which the scholar is also enmeshed. Over the years of research that have informed this book, my own job as a scholar and educator at the university has become increasingly intertwined with the sites, ideals, and practices I studied—including the less idealistic, more instrumental uses to which making has been put. During this period, universities sprouted incubators, makerspaces, and design thinking labs that promised to train both students and faculty in entrepreneurial skill sets and innovation thinking. When I started out as an assistant professor at the University of Michigan, my school set up an entrepreneurship program that employed the tools and techniques of making, acceleration, and design thinking in order to graduate "valuable employees," trained "to identify problems" on their own (UMSI entrepreneurship program). Similar efforts are nearly ubiquitous in universities, libraries, and educational programs across regions. And, as is becoming painfully evident, universities are not spared from the creep of neoliberalism.[127] In the United States and China alike, scholars are called upon to become entrepreneurial, to accelerate their output, to increase the numbers of citations they receive, to teach students to upgrade themselves into attractive human capital that serves the needs of the market economy. This feminist ethnography will thus frequently take us to scholarly and educational sites deeply complicit in producing the displacements of technological promises I examine here.

In-depth research on China's design, technology, and entrepreneurship cultures and histories remains sparse, with a few notable exceptions.[128] This book aims to further intervene in this field by teasing out the various threads that make up contemporary sites of technology production in China. These sites are rarely considered with the same careful focus or given the same countercultural cachet as similar sites and practices in the West; and rarely is scholarship that theorizes from the non-West taken up for its analytical and theoretical contributions alone. Similar to many of the people I met in my research in China, I have struggled with the relentless interrogation of what was the "uniquely Chinese" approach to making. Research in China is often interpreted as being about what is going on *in* China, but seldom read as having much to say about Western imperialism or processes of globalization. Research on technology practice in China is seldom taken up to inform theories of computing more broadly and beyond insights specific to China. Technology practice and research in China are largely construed as standing for a local and a national particular, despite the push back of a lineage of feminist anthropology that has introduced us to global assemblages, frictions, and political projects of decolonization.[129] This book shows that it is exactly such portrayals of China as incapable of defining anything of the scale as theory and modernity that enable colonial reverberations and their violence, legitimizing experiments with various forms of neoliberal restructuring and various forms of exploitation, within and beyond China's national borders. It is this tangled cat's cradle, woven of China's long-held aspirations of modern and global belonging, the proliferation of fears about increasing labor precarity and loss of agency in the West, and experiments in technological tinkering, that this book aims to loosen up.

Chapter Overview

The chapters that follow do not map a linear, temporal trajectory. Instead, they move between various temporal and spatial scales, pause at expressions of ambivalence, fall into gaps that interrupt smooth global connection, and move along the cycles of enduring promise.

Chapter 2, "Prototype Citizen: Colonial Durabilities in Technology Innovation" covers the years 2007 through 2011, when a collective of Chinese artists, designers, engineers, entrepreneurs, and Internet bloggers began to experiment with the ideals of the American free culture movement, participatory design, and eventually, open source hardware and making. Their work of transplanting Western ideals of participatory, open, and democratized technology production into contemporary China created an affective connection between China's history of manufacturing and its future as a global economic power. Their early experiments with participatory design, coworking spaces, makerspaces, and open source, open innovation, and open design, were aimed at prototyping a "new" Chinese citizen, i.e., the utilization of technology to cultivate an optimistic, forward-looking, entrepreneurial Chinese citizen, at last freed from connotations of lack and low quality. These attachments to technological promise are deeply intertwined with China's ambivalent relationship to the West, marked both by histories of colonialism and by revolutionary imaginings of alternative modernities.

Chapter 3, "Inventing Shenzhen: How the Copy Became the Prototype, or: How China Out-Wested the West and Saved Modernity," shows how a growing distrust of modernist ideals of progress in the broader tech and design imagination following the financial crisis of 2007–2008 found expression in shifting engagements with China. Specifically, I focus on how a series of influential actors, entangled with Western networks of investment, open source hardware, the arts, and design, turned to Shenzhen to make sense of technology's broken promises and partially redeem them. They portrayed China through colonial tropes of othering, framing its alleged backwardness, its associations with fake and copycat, as an opportunity space, to be celebrated for its difference. As technology was increasingly disassociated from its promise of modern progress, a new concern arose—that of the ethics and morality of the designer and engineer. As this chapter shows, Shenzhen was framed as a laboratory of exuberant scale that enabled the prototyping of a moral designer and engineer self who could claim to recuperate technological promise.

Chapter 4, "Incubating Human Capital: Market Devices of Finance Capitalism," documents how the venture capital system captures yearnings for technological alternatives. I follow the workings of a foreign-funded hardware incubator program in Shenzhen that trained people to translate their commitments to social justice into a pitch for finance capital. The incubator program taught people how to see themselves as human capital.[130] The chapter shows that the turning of the self into human capital is all but an inevitable outcome of neoliberal capitalism but has to be actively taught and learned.

Chapter 5, "Seeing Like a Peer: Happiness Labor and the Microworld of Innovation," builds directly on the previous chapter. It shows that entrepreneurial life required nurturing and maintenance. Specifically, I examine the exploitation of "happiness labor," the work of emotionally supporting precarious entrepreneurial life. Although happiness workers are often highly educated, happiness labor itself is low paid. These overqualified people are drawn in by the implicit promise that one day the happiness worker will be one of the entrepreneurs. Happiness labor is performed primarily by women and racial minorities. This chapter examines how the misogyny that is seemingly baked into tech bro culture in Silicon Valley was re-legitimized in China via the promise of peer production and openness.

Chapter 6, "China's Entrepreneurial Factory: The Violence of Happiness," documents the appropriation of making by the CCP, which manifested in infrastructural and urban upgrades aimed at inducing desires for self-transformation in Chinese citizens. I draw, here, from ethnographic research I conducted with the workers, designers, and managers in Shenzhen's manufacturing industry, documenting their ambivalent relationship to the "invention" of Shenzhen as a maker city. The chapter shows that by positing certain urban spaces as sites of making and as the ideal training ground for happiness and self-investment, the government aimed to induce in the workers, designers, and engineers of

China's manufacturing industry desires to upgrade themselves. It documents the slow violence at the heart of contemporary displacements of technological promise.

The conclusion returns to the labor and sites that have long challenged the inevitability of technological progress and its violence. It argues that if we attend to the labor that's necessary to nurture and sustain entrepreneurial life, we can mobilize other feelings to subvert the political economy of affect that runs on the promise of happiness. We can subvert the seemingly endlessly spiraling displacement of technological promise if we reframe what counts as intervention by moving away from our ideal types of countercultural heroism.

2

Prototype Citizen

COLONIAL DURABILITIES IN TECHNOLOGY INNOVATION

> People can inhabit enduring colonial conditions that are intimately interlaced with a "postcolonial condition" that speaks in the language of rights, recognitions, and choices that enter and recede from the conditions of duress that shape the life worlds we differently inhabit.
>
> —ANN LAURA STOLER, *DURESS*, P. 33

China's first coworking space, XinDanWei, was located in the western part of Shanghai, off Dingxi Road—a busy street filled with tiny restaurants, shops, and street vendors that sold everything from vegetables and gigantic plush bears to barbecued meat and fish. Amidst the buzz of neighborhood life, it was easy to miss the iron gate leading to the former rice factory that housed the coworking space. Once through the green iron gate, visitors found themselves in a quiet alley, wide enough for a truck to pass through. The alley was lined with grey walls on each side that were hidden behind a curtain of bamboo trees; water floated underneath, swallowing up the noise from the street. At the end of the alley was a courtyard, and at the back of the courtyard was a five-story building that stood out in this neighborhood of two-story lane houses. A line of large windows stretched over one wall. The rest of the walls, painted white and grey, were partially covered by a designer-like construct of climbing red ledges and stilts that gave the old rice factory a vibe of industrial chic. A large freight elevator transported people to the offices upstairs—XinDanWei, an architecture bureau, a photo studio, and a fashion magazine, all renting the space from the Shanghai government.

This part of the city, Shanghai's former French concession, was an urban reminder of what the CCP refers to as China's "period of national humiliation"—the foreign occupation and colonization of parts of China in the nineteenth and twentieth centuries. In the 1990s and 2000s, the CCP began to rewrite this humiliating past by conferring on Shanghai, a "once-and-now-again global city," a kind of "special modernity."[1] As part of the CCP's attempt to revise China's global image (and its self-image), the nation's five-year plan stated goals to build up China's creative industry (中国创意产业 zhongguo chuangyi chanye), with Shanghai as its global representative. The Shanghai government's own 11th five-year plan officially codified creativity as a central plank for development. Its old city neighborhoods were redesigned to become creative industry clusters. Old factories, lane houses, and at times whole districts were remodeled into trendy studios and office spaces like XinDanWei's building, a factory that was converted to a "creative industry hub" in 2004. The urban redesign made Shanghai a UNESCO City of Design—a designation it shared with such cities as London, New York, Paris, and Tokyo—and placed it into a global "creative city network."

Government discourse from the time framed the urban redesign in terms of social renewal. China's humiliating colonial past and the endurance of Western imperialism (e.g., in the form of American *cultural* hegemony) could be overcome, the CCP argued, if China's people became innovators who harnessed China's unique *cultural* characteristics for national development. The moral imperative to self-upgrade had been at the heart of China's economic development strategy since the 1980s opening reforms. Following China's entry into the WTO in 2001, the government framed this project of advancing national economic development as hinging on the self-cultivation of Chinese citizens as producers of "indigenous innovation" or "self-directed innovation" (自主创新 zizhu chuangxin).[2] This national branding strategy aimed both outward, working to rebrand "'made in China' as a trustworthy and affordable brand for the global consumer," and inward, seeking to project a "developmentalist vision" for China that would "assure its citizens of the nation's future and the role of the state in shaping that future."[3] The government promoted these social, urban, and economic upgrades as a project of overcoming China's reliance on manufacturing (and by extension as a project of overcoming China's reliance on foreign investment and outsourcing) by building a "self-directed" and "self-reliant"[4] innovation and knowledge economy."[5] Liu Shifa, then minister of culture, put it this way in 2004: "Contemporary China should be a creative China; from manufacturing to creative work, from 'made in China' to 'created in China.'"[6] The minister here echoed President Hu Jintao himself, who had declared "culture" as a "factor of growing significance in the competition in overall national strength . . . as part of the soft power of our country to better guarantee the people's basic rights and interests." Products of the culture industry such as films, the arts, animations, and magazines alongside the buildup of an indigenous tech and design industry should help increase China's global soft power. All of this was portrayed to be in the interest of the people—their self-upgrade into innovative producers would enable what was good for them: China's economic development, China's soft power, and social stability.[7]

It was at the coworking space XinDanWei, located at the heart of the urban renewal project aimed at enabling exactly this political project of building China's indigenous innovation economy,[8] that an eclectic group of people began gathering in 2009. They experimented with an array of ideas from the visions of the American free culture movement, DIY (do it yourself) making, and open source software and hardware tinkering to Chinese philosophy and craftsmanship. XinDanWei was the place where China's first hackerspace opened, where China's early Internet bloggers got together, and where both Chinese and foreign freelancers and entrepreneurs lived the promise of the creative class and flexible work by renting desks on an hourly, weekly, or monthly basis. All of this felt—to the people involved—radical. The ideas of a free and open Internet, of flexible work, and of the creative self seemed at odds with Chinese society labeled, both nationally and abroad, as risk-averse and hierarchical. It also seemed at odds with a government known for its tight grip on free expression. A feeling of excitement surrounded their project. The idea that technological experimentation might enable the buildup of a more open and democratic society in China was exhilarating.

But why would people who are invested in building an open society and a free Internet rent office space from a government that was opposed to such ideals? Did the experiments in China's early coworking spaces and hackerspaces challenge or enable the political project of creating China's self-governed innovation economy?

In this chapter, I show how the desire to achieve parity with the West (induced by the endurance of colonial frames of China as backwards, as the West's copy and "other") brought the CCP and the grassroots tinkerers of China's early hacker- and coworking spaces into a paradoxical and often highly ambivalent alignment in their respective projects to assert China as innovative and creative. I unpack how the very framing of certain technological practices as "grassroots," as open, and as providing agency to the people occludes the violence and the precarious conditions of life these seemingly promising technologies of "new" work proliferate.

The people who set up China's first open source hardware companies, who organized start-up weekends, hackathons, BarCamps, maker and TED events, had come together to contest the construal of China and its people as copycats, as low quality and as lacking the capacity to innovate. They were drawn to the free culture movement, to open source hacking and DIY making, to coworking and flexible work, because all of these technological experiments with work and life itself had been "pitched" by prominent and widely admired hackers, makers, artists, and investors of established innovation hubs of the West (from Silicon Valley to New York and Berlin) as providing agency to individuals (amidst corporate and state control). They had promised that self-actualization and individual empowerment via technological tinkering was available to "everyone." In the late 2000s, these "new" models of tech work and innovation from makerspaces to coworking spaces and tech incubators were portrayed as promising to intervene in established corporate and political structures. They were celebrated as democratizing innovation beyond closed research offices and design labs by providing individuals with the ability to start their companies, build their own jobs, and make their own technologies. Advocates of these "new" models argued that by democratizing tech innovation, social justice would be achieved. This socialist pitch portrayed self-economization as providing agency (it portrayed economic life as entrepreneurial life). It enabled the endurance of an "old" technological promise—the kind of techno-optimistic ideal promoted in the 1990s by America's free culture advocates who had argued that—as scholar of communication Fred Turner shows it in astute historical detail—"social change would come not from political mobilization" but from "experimentation and the proper deployment of the right technologies."[9] Coworking spaces, makerspaces, and incubators were positioned to democratize this technological promise of agency, liberation, and individual freedom. They further cemented the idea that social change was to be facilitated by technological experimentation and not by political intervention—a depoliticizing move. The entrepreneurial (and seemingly apolitical) life that America's free culture advocates had rendered desirable in the 1990s suddenly appeared to be in reach for "everyone"—even for people in China long deemed (by the CCP and the West alike) "bad pirates" and lacking the capacity to be innovators.[10] In this chapter, I show how flexible work, open source, and venture labor all produced a feeling of optimism and agency in China, exactly because they had been "pitched" as bringing about the good life, as providing justice and agency for Chinese people.

It was this depoliticizing move (at the heart of the socialist pitch) that aligned experiments in open source hardware, flexible work, and a free and open Internet with the CCP's own political ambition to build an indigenous innovation economy. Simply put, what else could have been more in the CCP's interest than supporting the idea that people should become technological tinkerers rather than political activists? The project of prototyping new ways of working and living in China by setting up coworking spaces, makerspaces, and incubators felt exciting to many people in China exactly because none of these initiatives were understood as a government directive. They felt grassroots, self-driven, and even countercultural. They were associated with

Western liberal ideals of individual freedom and empowerment rather than with Chinese society and state control. It was because of this affect of intervention attached to these "new" technological experiments that their official government endorsements seemed exciting to many. When the CCP posited China's makerspaces as incubators for the people and their founders as *prototype citizens*, i.e., citizens who modeled "the good life" via technological experimentation for other Chinese, many agreed that this was a good thing—the government seemed to endorse their own desires to contest innovation and remake China from copycat into innovator. The promise that Chinese people too would be recognized as creative makers and innovators (both by the West and the CCP) made people endure injustice and continuous colonial tropes of "othering." This promise of being (at last) granted the status of happy, modern innovator occluded the proliferation and normalization of precarious conditions that these "new" ways of working brought with them and the CCP's own neoliberal agenda to make its citizens responsible for their own happiness, livelihoods, and futures.

Contesting Innovation, Contesting Modernity

> People say that Chinese have no creativity. That's bullshit! There are lots of very great ideas, some of them are almost too incredible to believe. We are not in shortage of people with good ideas. What we lack are the ability to execute, to extend and the power of influence and resources. Where can you get those things? . . . if there is a place where people can meet each other and contact all those resources, what will happen then? That's why we started XinDanWei.
> —LIU YAN, TEDX SHANGHAI, 2010

Picture yourself back in Shanghai, where we began, in the courtyard of the refurbished factory building where the coworking space XinDanWei was located. It is July 2010 and just a couple of months ways from the Shanghai Expo—the World Expo that cemented Shanghai's status as a world-class city and China's rising soft power.[11] You stand in front of the five-story building and press the elevator's button for the top floor. When the elevator doors open, you see a woman in her late fifties seated on an old stool, its top covered in what once must have been a vibrant red velvet, next to the elevator's operating panel. "几楼 jilou—Which floor?" she asks firmly, ready to push one of the buttons. As the elevator begins to move, you realize that it is larger than you first thought, the marks of heavy-duty machinery visibly engrained in its floor and side walls, hinting at the building's former use. You notice a sign on one of the walls. In big white letters on a blue background, it says "The Human Side of Work." Underneath, there is a series of photographs: the interior of an office space, a cactus adorned with a Creative Commons sticker, a young man writing on a wall covered in black chalkboard paint, and a group of people gathered around colorful materials spread across a clean, white surface. "XinDanWei facilitates creativity, sharing, and collaboration," and "this is a new way of working," you read. A pink balloon and black umbrella partially cover the right corner of the sign. The balloon floats up, hit by a shaft of air from the old rotating fan mounted to the elevator ceiling, and you suddenly see another sign that says in Chinese, "civilized (文明 wenming) behavior is paramount." This sign shows a childlike cartoon of three Chinese citizens being reprimanded by police for inappropriate public behavior—spitting on the subway, cutting in line, yelling. Before you finish reading, the woman in the elevator announces loudly, "到了 (dao le)—(we) arrived!" You step out and find yourself in front of a glass door through which you see an open-office design with desks dispersed throughout, a cactus plant adorned with a Creative Commons sticker, and a wall painted in black chalkboard—XinDanWei, a place where the significance of the police ordinance you just saw fades into the background.

In the months leading up to the 2010 Shanghai Expo, such ordinance signs were an unremarkable sight in Shanghai. They were mounted on escalators and elevators all over the city, in housing compounds, subway stations, schools, libraries, and universities. The juxtaposition of this particular ordinance sign with XinDanWei's poster promoting a "new human" that was enabled by "new ways of working" might have been serendipitous, but the connection between the two adjacent posters reflects something real. Since China's entry into the WTO, the Chinese national project has aimed at combining an often Western-identified narrative about creativity and technology with official state discourse about Chinese citizens' responsibilities and capabilities, rooted in claims of people's lack of "quality" (素质 suzhi) and of "civility" (文明 wenming). The XinDanWei coworking space and the people it had gathered since 2009 were central to the crafting of an alternative narrative about the creativity of Chinese people that operated alongside (and was envisioned to infiltrate) this official rhetoric. XinDanWei had become a site where entrenched understandings of innovation and creativity, be they Western or articulated by the CCP, could be openly negotiated and contested.

During my research with XinDanWei in 2010–2012, it was becoming an increasingly thriving place where people who understood themselves as members of a transnational tech and creative industry network gathered for small- to large-scale events: China's early Internet blogging conferences, BarCamps and Dorkbots, TEDx conferences, creative industry and smart city workshops, hackathons, research lectures, informal debates, weekly "chit chats,"[12] and art exhibitions. Many of these events were hosted at XinDanWei or had one of the XinDanWei co-founders as organizers or co-hosts. These events were attended by two main groups of people: fairly affluent young Chinese (ranging in age from early twenties to late forties), many of whom had studied and/or worked

FIGURE 2.1. 2010 Shanghai Expo poster promotes 精彩世博，文明先行 ("Wonderful World Expo. Advance a Civilized Culture."). Just a bit farther down on the same street, a door marked with the Chinese character "chai" (拆) indicates that the two-story low-income housing compound behind the wall is slated for demolition. Photograph by the author.

abroad, and foreigners, some of whom worked in Shanghai's international corporate tech industry (at companies such as IDEO, Frog, Google, Microsoft, and Intel) and others who freelanced or consulted for Chinese and international architecture bureaus, design firms, and investment firms. As international attention to XinDanWei grew, they entered into a series of informal partnerships with people at American and European institutions such as Creative Commons, the Berkman Klein Center, TED, and THINK. These partnerships grew largely out of the co-founders' own personal and professional networks.

The people they invited to give lectures or show their art and design work, from China and abroad, the readings that circulated in and through the activities at the coworking space, and the writings that some of XinDanWei's core members produced—albeit varied in their motivations and ambitions—shared an underlying ideal. It was the ideal that if you empowered people to build and design their own technologies, their own organizational structures and forms of tech work, they would be empowered to also enable broader societal change. They were not focused on "teaching" people how to be creative, but on reformulating the very relationship between the individual, the state, and the nation—they were committed to prototyping alternatives to official frames of what constituted the good life, including citizenship. These efforts to assert Chinese people as autonomous selves (freed from colonial tropes of lack such as wenming) would later appeal to the CCP in the very moment it required its citizens to prototype the Chinese nation as a worldly, cosmopolitan, and global entity (see chapter 6).

The discourse of wenming first emerged during the Opium Wars, when it became—as the historian Andrew Jones describes it—"an emblem of all that was advanced, standing as a synecdoche for the power and prestige of the West, and marking the geographical and historical rupture between here and there, old and new. It signified, for all intents and purposes, the modern, insofar as that modernity was inescapably inscribed within a colonial relation to the West. It was, in this sense, a deictic term, one that pointed not only to a thing in itself, but to the relation between self and other."[13] During the 1980s and '90s economic reforms, the CCP employed this discourse of lack to induce desires for self-upgrade. It portrayed China's citizens as being of low "quality" (素质 suzhi) and positioned China's inability to modernize as a failure of the Chinese people to embody international standards of modernity, discipline, and civility (文明 wenming). As anthropologist Ann Anagnost notes, the notion of wenming came to stand for many things: modernity, westernization, civilization as an advanced stage of historical development, international standards of labor discipline, orderly behavior in public places, and more.[14] Most important, though, wenming was portrayed "as the inescapable reality of a new world order, a necessitated development" for China to not necessarily become like the West, but represent an "advanced culture" in international comparison.[15]

China's future, its reputation, and its economic development were portrayed as hinging on the cultivation of a particular kind of citizen—one who would associate self-improvement (and improvement in their personal social and economic status) with the improvement of the national economy.[16] The citizen would reflect the nation's reputation.[17] Consumption and self-fashioning were positioned not only as a source of personal happiness and transformation but also as a way to attain a certain global status for the nation—what anthropologist Lisa Rofel (building on Michel Foucault) calls a "postsocialist technology of the self."[18] The socialist experimentation and class struggle of the Mao era were replaced by the neoliberal "language of management."[19] This discourse of optimization—of life, expertise, productivity, and the self—was designed to cultivate "a new human" who would usher China into a new era. Modernity was to be achieved not through socialism but through "structural adjustments of the market and the self."[20] The socialist state gave way to the "enabling state," fostering citizens who understand themselves in terms of their value to the nation, their value as human capital (referred to as rencai 人才 in Chinese).[21] The citizen, in other words, became construed as an asset for the nation's project of future making.[22]

With the aforementioned creative industry policies of the early to mid-2000s, the state aimed to transform China's past, its culture, and traditions into fuel for China's economic development. The goal was to position China to compete in a global market that views culture as a resource for value creation. To achieve this aim, the CCP continued to demand that citizens self-upgrade to forward national interest. And it continued to draw on the civilization discourse of wenming and suzhi to integrate citizens into "the dream of regaining China's stature as an empire"—driven by desire to "attain material and moral parity with the West."[23] This ideological "practice of nation branding," scholar of communication Fan Yang shows, "privileges the national over the social," with the citizen construed as serving the redemption and competitiveness of the nation on a global stage.[24] In other words, suzhi and wenming discourse have positioned the citizen—as scholar of communication Cara Wallis elucidates— as "hyperfeminized" (subordinate, self-sacrificing, lacking a political voice) and the state as masculine (in pursuit of economic development).[25] These discourses thus reveal the party's ambivalent stance toward China's people; as Anagnost argues, "the people" are both what is desired—rendered as crucial for the making of the nation—and at the same time "what is feared," what is "held at a distance by a consistent disavowal of their exclusion from the political process."[26] Despite the state's ambitions to disassociate its own modernization from the West, the uptake of creativity by the CCP since 2001 ironically legitimized, Yang explains, a globalized hegemonic discourse of creativity and innovation, rooted in Western-centric modernist ideals of individualized authorship and novelty.[27] Chinese citizens were called upon to self-transform into the kinds of innovators who file patents, work according to the standards of IP and copyright, and advance Chinese culture into assets for the soft power of the nation. The kind of technological experiments that unfolded at XinDanWei were driven by desires to intervene exactly there—the relentless colonial othering of Chinese people that both their own government and the West upheld.

XinDanWei began as a series of informal gatherings in Shanghai between 2008 and 2009. Two of the three co-founders, Liu Yan and Chen Xu, had just returned from years of studying and working in Europe, Liu Yan in the Netherlands and Chen Xu in the UK. After their return, the two women began working with the electronic media artist aaajiao, who had just begun making a name for himself in the international electronic arts scene. Together, they hosted SanShu (三书—three book) Salon, which were ad hoc monthly gatherings in coffeehouses, art galleries, libraries, private homes. SanShu Salon brought together people from their transnational network of friends and colleagues active across the fields of the arts, technology, and research. The network grew, and in 2009, Chen Xu, Liu Yan, and aaajiao opened the coworking space XinDanWei. Like other coworking spaces, XinDanWei was established as a shared community space that provided desks and rooms for people to rent on an hourly, weekly, monthly, or yearly basis. For the three co-founders, coworking was about more than renting out office space in a cool location. For them, it constituted a crucial "platform" that would enable "other Chinese people" to experiment with what they called "new ways of working"—such as starting your own business, consulting work, artistic and creative practice, or mobile, flexible work outside of what they considered "traditional" structures such as large corporate offices, research laboratories, or state-owned enterprises. But it was about more than simply motivating people to become entrepreneurial. It was about creating an environment that would breed a particular kind of entrepreneurial attitude, one based around the idea that an open approach to design, engineering, and technology empowered individuals to define on their own how they configured their relationship to society, the economy, the state, and the world writ large—it was the idea that people can and in fact should prototype alternative models of being (including citizenship). They called it a "DIY (do it yourself) approach," which they differentiated from the kind of top-down demand for creative production they associated with the government and with Chinese education. They believed that this form of creativity, which built on each individual's capacity to experiment and play, would lead to drastic shifts in Chinese society. "All of this is not a business with huge profit," Chen Xu explained to a group of researchers from Intel when they visited XinDanWei in 2011, "It's more about a turn of society as a whole, where creativity and innovation are more and more appreciated. Our coworking space brings together all the creative energy we see in China today and turns them into another level."

As documented by anthropologist Lily Chumley, art and design education at elite institutions like CAFA (Central Academy of Fine Arts) in Beijing and in Chinese colleges more broadly was framed by the CCP in the 1990s and 2000s as the ideal training ground for the cultivation of rencai (human talent/capital) and the cultivation of creativity among China's people.[28] The aim of art and design education was to mass produce "'civilized' (wenming) citizens, with 'high' suzhi or value—the qualities of being that characterize educated, urban, middle-class subjects."[29] As Chumley shows, these schools produced young people with "neoliberal agency" who were trained as "prototypic liberal individuals" within an otherwise "illiberal state."[30] In other words, these citizens were stipulated to be prototypes—trained to model the transformation desired by the CCP: a transformation from socialist citizens who were focused on politics and class struggle into neoliberal citizen subjects who were focused on economic self-fashioning.

The collective of people around XinDanWei had come together in their desire to "truly" "open up China" and empower individuals as "truly" self-defining agents freed from both Chinese and Western conceptions of citizenship. They shared a belief that this empowerment could not happen inside elite institutions like CAFA, critiquing China's existing educational structures as highly bureaucratic, top-down forms of learning that undermined the true creative potential that "all Chinese" carried. They saw themselves as both modeling this DIY version of innovative creativity and providing the kind of open and collaborative institutional frames and platforms that would enable "all kind of Chinese people" to see themselves in a new way. They aimed to free Chinese people from the pressure to perform an already defined notion of citizenship—from family pressure to succeed in China's competitive education market and obtain a good job and from societal pressure to embody a particular form of creative agency and be a "quality" citizen. The key ambition of the people who had gathered around XinDanWei was to cultivate a self-empowering version of the prototype citizen, one that would not only "model" the good life but redefine it and experiment with possible alternatives to existing norms and approaches. And it was because of these yearnings for an alternate conception of Chineseness, one not associated with lacking creativity and civility, that people began gravitating toward the socialist pitch of entrepreneurial life that this very interventionist agency so longed for.

Yearnings for Justice

In 2010, Liu Yan, the CEO of XinDanWei, was invited to submit an application to the "Girl 2.0" competition in Shanghai. Of the three co-founders of XinDanWei, Liu Yan was the most publicly known, serving as the face of the coworking space. She was regularly invited to give talks at local, national, and international tech and creative industry gatherings; she wrote blog posts and research publications about coworking and open collaboration; and she consulted for Western firms from IDEO to Philips on fostering open collaboration and innovation in their China facilities. The day she received the invitation to Girl 2.0, she had returned to XinDanWei from one of these events. "Look what I got," she said as she sat down next to me and Paris, the twenty-three-year-old intern and designer who had been hired in early 2010 to help XinDanWei with corporate branding, marketing, website design, and space design. "It's an invitation to submit to this Girl 2.0 competition." Paris and I took a step closer to look at the flyer in Liu Yan's hands. "Paradigm shifts: the making of Girl 2.0," the flyer began. "What started as a mission to empower women in technology transformed into a vision of shifting the paradigm of global female innovation. Come hear the story of how Girl 2.0 is building an ecosystem that is changing the way women work with and support one another."

Liu Yan had met one of the women who was managing the competition at a small gathering of international creative industry enthusiasts. "She told me XinDanWei is perfect for this. They want to focus on innovative companies found or led by women," Liu Yan explained. She laughed at the language of the flyer in her hand. "It's a bit much," she said, "but you know, it would be a good chance to promote XinDanWei and give it some visibility in the creative industry scene." "Let's do it," Chen Xu chimed in, overhearing Liu Yan's last words as she left a meeting in the glassed-in room next to us. "Silvia, you have some experience with video production, right?" Liu Yan asked. "They want a video submission, it seems." I confirmed that I had indeed done video production in college; although I was a bit rusty, I'd be happy to help. The submission deadline was in two weeks, so over the following days, Liu Yan, Chen Xu, Paris, and I worked feverishly on a storyboard. After a session of hammering out the crucial takeaways about XinDanWei that people should have after the video, Liu Yan declared,

> The key is that we demonstrate that at XinDanWei, what we do is innovation based on sharing. And we have to demonstrate that this is different from any other platform of working. We have to show that XinDanWei means a new way of working. And that this is very simple and not complicated to do—that everyone can do it. The film should also show that we are an open network, rather than a closed society. XinDanWei is the future work style for China. This is what is creative about our work. And this is what we are contributing to China. At XinDanWei, we cultivate new ideas. China means creativity and creative people. So we are creating a work environment to stimulate the cultivation of new ideas and creativity.

The film's aesthetics and its very mode of production, everyone agreed, should convey XinDanWei's core commitments. Crucially, it should show that

FIGURE 2.2. XinDanWei film production, 2010. Photograph by the author.

creativity was something that was already present in Chinese society, and that it could be unleashed through the principles of collaboration, open sharing, and peer production. The production of the film itself demonstrated how this could work in practice; as Liu Yan explained it, "the making of the film will show what is possible with collaboration and open sharing," for the production of the film mobilized people in XinDanWei's network to help produce it. One member offered the free use of his professional photography studio during the evening hours, another lent her professional film camera, and still others helped with set design or served as actors. After only four days of work, mostly during people's off-hours and evenings, the film was shot, cut, and released on a series of online video sharing platforms.

The film and its production reflect the affect of entrepreneurial living. Many felt that if people were able to change their way of working—to become, for example, self-employed geeks, electronic media artists, or start-up entrepreneurs—all of Chinese society would become more "open," and citizens would have the power to shape society in ways potentially different from the nation-state's interests. According to the co-founders, all of XinDanWei's public-facing activities—day-to-day operations, media promotions (such as the short film or the website), space design, and naming practices—were designed to signal to "other Chinese people" that "there was no harm in opening up," and that it was "simple" to do, fun, and rewarding. The space's modern, minimalist website, with its black Helvetica font on simple white background, proclaimed the company's values: "XinDanWei stands for Simple, Open, Networked," and reminders of these values often circulated throughout the coworking space, written on cardboard signs placed on little stools or on the chalkboard wall at the entrance. The co-working space's naming practices and space design conveyed that these core values were not Western but "Chinese," and they encouraged Chinese people to "experiment" with these values, to try out what it meant to live them.

The name chosen for the coworking space reflects this aspiration, and points both to the future and to China's past. The name XinDanWei is comprised of the Chinese characters "Xin" (新—New) and "DanWei" (单位—workunit)—read together as "New Workunit." The danwei, a relic of China's socialist era, refers to state-owned institutions that functioned as the core organizing principle of social transformation after 1949. The danwei broke up large urban populations into smaller collectives that provided employment, housing, and social benefits for workers and their families.[31] They were essentially "housing compounds, constructed, owned and regulated by work units, which acted as de facto landlords and managers."[32] By dividing up China's large urban population, the danwei was a crucial element of the city's material and social fabric and an essential aspect of people's day-to-day realities as they structured work and family life under socialism. The repurposing of the name "danwei" was not intended to make a case for bringing back the organizational structures under socialism. The combination of "new" (xin) with "danwei" asserted that the socialist promise of collective living and organizing, embodied in the danwei, could be translated into a modern, digitally mediated form of collaboration that (unlike the original socialist danwei) would not deny individuals the opportunity for self-actualization, self-definition, and self-entrepreneurship. Indeed, the making of a better, more just society would thrive on precisely what had been denied under Mao's socialism: individual technological and self-entrepreneurial agency. China's socialist past, in other words, was invoked to demonstrate that there was something uniquely Chinese about this coworking space, that it was not simply a copy of coworking spaces elsewhere. It was the idea that China's socialist past—updated via the techniques of entrepreneurial life—had something to offer to China and the world.

The clever choice of the name XinDanWei signals the collective's broader efforts to test out possible alignments of the values and workings of Chinese society with those that might be at first considered antithetical. This project of harnessing something unique from China's past, culture, and society in order to assert Chinese people as creative agents on a global stage wove through the day-to-day management and design of the coworking space, from the choice of furniture to wall paint and office rentals.

"Let me show you how we decided on the colors for our space," Paris said to me on a quiet day in the coworking space. "See, I did an Internet search of the old danweis and found imagery of how they looked and the colors they used. The aesthetics actually have a lot in common with what we value at XinDanWei. See, these colors, they are simple, plain, white, and they have this basic furniture. There is something functional about them." He explained how XinDanWei's website and spatial design and layout reflected and updated these aesthetics, giving them a slight modern twist that made them contemporary and "cool." This work to align Chinese society's past with its future was central to the daily operations of the coworking space. The space design itself was considered key to enabling people to experiment with ways of living that might depart from societal norms or political pressures.

When XinDanWei moved out of the creative industry cluster described earlier in the fall 2010 and into a three-story lane house (a building typical of the city districts in the former French concession), the co-founders positioned the move as an opportunity to advance their experimentation with these alternative forms and practices of living and working available to Chinese society. "We want to get some new furniture," Chen Xu told me a couple of days before the move, "to really go with the danwei theme." aaajiao had mobilized one of his friends in the furniture business, and a couple of days later the new space was equipped with twenty vintage "iron pipe" chairs, of the type used in Shanghai office spaces in the 1970s. Over the following weeks and months, Chen Xu and Paris carefully placed and arranged a variety of artifacts in the new space. Chen Xu had found on Taobao an old punch card system that had once clocked workers' hours in a factory, and this was placed next to the entrance. Posters with drawings in the style of the Soviet era and Russian language annotations hung on one side of the downstairs wall space. Next to the posters was a collection of paintings by young female Chinese artists. Nearby stood a desk covered in Arduino boards and a 3D printer; beside the desk was a shelf that held a wide variety of Chinese and English books and magazines, including *Urban China*, *Wired* magazine, and Tim Brown's *Change by Design*.

Such attempts to recuperate China by refashioning its past—by reframing and aligning it with what was labeled "new" and innovative—were common to the various projects of China's early coworking and makerspaces. What fueled these projects of recuperating China were the continuous claims made by many of the foreign (mostly Western) workers (often employed by international tech corporations and design consultancies that had opened up shop in cities like Shanghai and Beijing) who circulated through China's coworking and makerspaces that "the Chinese" (still) didn't know how to be innovative and that it was people's attachment to stability and hierarchy that held them and China back. These contestations over what counts as innovative were common in China's fledgling tech and creative industries at that time. They were part of a discursive milieu through which municipal and district-level governments justified drastic urban development projects to attract "creative minds," while displacing those deemed incapable to self-upgrade into innovative talent.[33] Shanghai was often posited as ideal for China's larger branding strategies, for it had long been "characterized by juxtapositions between eras (where past and present met) and by juxtaposition between cultures (a place where East again met West)."[34] The city's urban redesign to build creative industry hubs and clusters like the one that first housed XinDanWei itself was branded as a success in rescripting the city's colonial past as an opportunity for the present. Shanghai demonstrated, in other words, how the cautionary tale of Western domination can become an asset for prototyping the nation anew.

At a 2010 TEDx event that I attended with Liu Yan, who was one of the invited speakers, I met Jack Ho,[35] the general manager of a development firm headquartered in Shanghai (and backed by a Hong Kong investment group). The firm had played a central role in the refashioning of Shanghai's past to assert global leadership in the present. Ho had joined the firm in 2004 after many years of working in Silicon Valley. Just a couple of years earlier, the firm had received an architecture innovation award for the successful remodeling of parts of Shanghai's former French concession into a district that housed museums, shops, restaurants, art galleries, and design boutiques. It was a project that turned the city's past into symbolic capital. Preservation served the making of global markets of consumption.[36] For Ho, the job in Shanghai meant that he would be able to "bring innovation to a new market"; he explained this to me in an interview a few weeks after our first meeting, in response to my question about why he had given up his job in Silicon Valley and moved to Shanghai.

> For me, this was a question of how to transfer these lifestyle choices and innovation ideas [of Silicon Valley] into China . . . Peter Drucker once said "innovation has to come to the market." I see Shanghai as very entrepreneurial; it does know how to bring innovation to the market. My other good friend Leo Lee wrote *Shanghai Modern*. It's about the speed to market and scale that's tremendous. Shanghai knows how to scale. We looked for a place with intellectual capital. I looked at over 200 cities in China. I could count on 2 hands to find those places that were suitable, based on their culture and ability . . . The vision is really—well, that's the reason why I came to China. I wanted to help a place to build a knowledge community.

The book Ho referenced here, Leo Lee's *Shanghai Modern*,[37] argues that it was the foreign presence in Shanghai in the 1920s and '30s and the kinds of public and social spaces (cinemas, coffeehouses, department stores, parks, dance houses, and so on) it had built that enabled the construction of modern Chinese cosmopolitan culture. From this viewpoint, modern cosmopolitan Shanghai "could be understood not as the cultural domination by the foreign, but as the appropriation by the local of 'elements of foreign culture to enrich a new national culture.'"[38] The colonial past, here, is framed as an opportunity— as what enabled Shanghai to "scale," to attract "intellectual capital" and to be "suitable" places to live for people like Ho.

At the time of my interview with him, Ho was working on precisely such a project of "national enrichment"; his firm had been leading the redevelopment of an area in northwest Shanghai (WuJiaoChang, in the YangPu district) into a Knowledge Innovation Center (KIC)—what would become a flagship project for the development firm. It was one of many urban development projects that rendered China's colonial history attractive to financial investment. The stated goal of the project, Ho explained to me, was to build a space that would foster "a new class of entrepreneurs, creative thinkers and innovators." Key to attracting the "right" kind of people to the KIC in the first place, Ho told me, was that the developer had identified the right location. The location chosen was an area that government officials had designated an economic and technological development zone in the years before the Japanese invasion. The development zone designation freed the area from a series of restrictions and regulations (for instance taxes and labor laws), opening it for re-development.[39] The KIC area, which bordered some of the city's major universities (Fudan and Tongji University) and was convenient to the city's expanding subway system, had the potential to attract what Ho called "innovative minds." He envisioned the KIC to work as a "resource platform" that would insert Shanghai into the global network of tech innovation. To this end, he planned the setup of a digital communication infrastructure that would allow occupants of the KIC to connect instantaneously to Silicon Valley via cutting-edge teleconferencing (to my knowledge, this digital infrastructure was never actually built). China and its people would be redeemed (and shed any lingering attachments to its colonial past) by providing the "right kind of" urban infrastructure that enabled people's self-upgrade into the innovative producers the nation could be proud of.

It was this desire to redeem Chinese society that brought into ambivalent alignment state actors, urban planners, and the people who had gathered in China's early coworking and makerspaces.

During a long subway ride following an event, Liu Yan told me a story from her time in Europe: "You know, when I lived in the Netherlands," she explained, "people always thought I was the one who ran the local Chinese restaurant. They just couldn't see me in any other way. They couldn't see Chinese people in any other way. It kind of pushed me, I guess, to work harder. I got this degree in arts management. I now hold a Dutch passport. And so I realized one day that I am in a position to bridge across these cultures. One of the first things I did, before I came back to China, was that I started a datong desk in the Netherlands."

For Liu Yan, "datong desk" denoted a "European–China alliance" aimed at "facilitating international collaborations between the creative industries in China and Europe." Datong desk was a play on datong (大同), a Chinese philosophical concept that merged Confucianist concepts with European enlightenment ideals.[40] It became understood as a Chinese utopian modernization theory when it was picked up by the Chinese reformist intellectual Kang Youwei around 1902; published as datongshu in 1911.[41] Datong emerged from an intellectual tradition of reformists eager to work toward a version of Chinese modernity that escaped the colonial world order; Kang Youwei's vision was that of a one-world government that considered all nations equal. Datong, for instance, has been variously translated as "great harmony," "Great Unity," "Grand Union," "Cosmopolitan Society," "Great Commonwealth," "the Great Similarity," "Great Universality," "Era of World Brotherhood," "Great Communion," "Great Similitude," "Great Similarity," "One-World Government," and "Cosmopolitanism." In the 1990s, the term was picked up by a group of Chinese intellectuals (and a few CCP members) as part of their attempt to reposition China as an emerging, peaceful power in international politics.[42] These groups argued that China could "offer an alternative . . . model of world order for the 21st century."[43] In a series of talks Liu Yan gave in Chinese, she took up this ideal of an alternative world order rooted in mutual respect and cross-cultural engagements:

> It's a little too much if I say it's imperialist invasion, but in fact when two countries try to exchange their culture, there are always many ideologies in it. It's not my goal to help people do that type of spreading their own culture. I want to build my own business. I like to have my own projects and enrich this kind of cross-culture, cross-discipline community. That's why I made datong desk. Datong is named from Kang You Wei's "Great Harmony in the whole world." Because I think actually we as human beings have the same universal language. We have the same language, there are only a few differences between us. We can cooperate in many ways; all differences can be mixed and crossed.[44]

Aspirations like these—to rescript China as a hopeful model, a prototype for alternative ways of living—weave throughout China's postcolonial history. The dream to redeem Chinese people as being seen (by the West) just as modern as Westerners has become inseparable from the party's political ambitions to redeem the nation as a sovereign self-defining entity and to assert itself as *the* only leadership capable of doing so. The uptake of venture labor, flexible work, coworking, and adjacent models of tech innovation promised to provide the means to achieve this much-longed-for *parity* with the West and justice for Chinese people.[45]

As I show in chapters 4–6, it was these technological promises of individual empowerment and "user participation" that would also channel yearnings for political change and justice into economic action; such technological promises create the feeling that when people turn themselves into technological tinkerers, flexible workers, and venture laborers political change will be possible—that people might achieve justice so long denied. It is this affect of intervention that masks the depoliticization that accompanies the displacement of technological promise—people become invested in technological and economic rather than political action.

The Promise of the Liberated Self

In the summer of 2010, XinDanWei hosted one of China's few early Internet Blogger conferences. A few days after the event, I noticed a new artifact placed on the shelf at the main entrance of the coworking space: right next to the award from a government-sponsored "creative industry" competition, a batch of Creative Commons stickers (one of which somebody had stuck on a cactus plant right next to the shelf), and promotional flyers for an open source robotics event sat a clock that I had not noticed before. Upon closer inspection I realized that the numbers that would typically indicate the clock's time were replaced by a small figure in the shape of a crab—a reference to what was celebrated at that time as China's homegrown Internet counterculture. The Chinese translation of crab—or rivercrab to be precise—is 河蟹 hexie. When Internet censorship was mounting in the mid- to late 2000s, it was this innocent Chinese character of the rivercrab that came to function as a tactic to critique censorship online, while circumventing censorship itself. Allow me to unpack briefly. With a slight shift in tonal pronunciation, hexie no longer translates into rivercrab, but into harmony (和谐 hexie). In the mid- to late 2000s, the Hu Jintao government had begun to legitimize Internet censorship by pointing to the rise of Internet addiction and social instability online.[46] It argued that state control of the Internet was necessary to guarantee the "healthy" and "harmonious" development of Chinese society.[47] Over time, to "harmonize" thus came to mean "to censor." As "harmony" increasingly became a synecdoche for "censorship," the usage of the Chinese characters for "harmony" became highly regulated online, and Chinese Internet users began substituting the characters for "rivercrab" (河蟹 hexie) in place of those for "harmony" (和谐 hexie) to avoid triggering the censorship algorithms.

FIGURE 2.3. Hexie (rivercrab) Clock at XinDanWei. Photograph by the author.

These creative and playful usages of the Chinese language, aimed at navigating around and critiquing censorship, were documented with great excitement by the Western press and scholars of the Chinese Internet. Many speculated that the Internet could be a liberating force, enabling Chinese citizens to become their "truer" selves, or what many began referring to as "netizen" (网民 wang min in Chinese). As the Internet and media scholar Hu Yong wrote on his blog at the time, "The social power of Internet technology is concentrated in the figure of the netizen," and the government must recognize the "networked power of Chinese netizens" in order to "leverage the Internet's productive force for economic and social development." The Hexie (rivercrab) Clock at the Xin-DanWei coworking space embodied this very belief that Internet technology by way of empowering Chinese citizens to become liberated and free individuals would also constitute a "productive force" of national, economic development.

The Hexie Clock, thus, was an artifact that symbolized a displacement of technological promise—the ideals that the kind of entrepreneurial life championed by America's free culture advocates and its subsequent incarnations was now possible in China. A feeling of subversive energy evaporated from artifacts like the rivercrab clock, co-produced by their eager Western celebratory endorsements and Chinese yearnings for parity and justice. Invoking such feelings of technological interventionist capacities brought international and especially Western legitimacy. One such person, who had successfully positioned himself as a spearhead leader of China's Internet counterculture, was the Chinese Internet blogger Isaac Mao (who was a frequent visitor at XinDanWei and also was the one who had gifted the Hexie Clock to XinDanWei). Mao had become widely known in international tech and electronic arts networks after a 2010 BBC interview in which he openly discussed how to circumvent Chinese Internet censorship. While this interview had political ramifications for Mao in China, it turned him—almost overnight—into a symbol of the Chinese Internet counterculture in the West.

I first met Isaac Mao when he was invited to speak at the 2010 Chinese Internet Research conference, which for the first time was located in China (Beijing). At the conference, Mao was treated as a star, with a large crowd of people surrounding him after his talk. He had just returned from a visiting position at the Berkman Klein Center in Boston. Over dinner, I sat next to him and he excitedly told me about a series of collaborations he had entered with Joi Ito, the then-chairman of the Bay Area–based non-profit organization Creative Commons.[48] Creative Commons was a key institution in proliferating the American ideals of the free culture movement. Creative Commons, founded by the lawyer Lawrence Lessig in 2002, is a non-profit organization that provides non-copyright licensing mechanisms. In his 1999 book *Code*, Lessig explains the vision that drove his development of Creative Commons: for Lessig, empowering technology users to become creators who control how their ideas are used would enable a commons, and this commons would in turn be the key to "limiting, or checking, certain forms of governmental control."[49] It was precisely Lessig's decision to portray Creative Commons as a form of technological intervention that carried potential for political intervention, as anthropologist Gabriella Coleman shows, that earned the project such broad legitimacy.[50] And it was exactly this socialist pitch—the promise that social justice and political intervention could be achieved via technological means while remaining seemingly apolitical—that Isaac Mao built on in his writings about China's Internet society, which he framed around the concept of "Sharism" (分享主义 fenxiang zhuyi in Chinese).

Mao positioned Sharism as an "ideology" of and for China's Internet age. He argued that the Internet, and in particular social media, would (if uncensored) create a societal "mind switch" and by extension benefit China's social and economic development. The idea of Sharism found traction in the Western open tech and electronic arts scenes. It was published, for instance, as a book chapter in an edited volume Joi Ito put together in 2008 (*Freesouls*) that also included contributions by Ito himself as well as other prominent figures in the scene: Lawrence Lessig, Christopher Adams, Howard Rheingold, Cory

Doctorow, Yochai Benkler, and Marko Athisaari. Mao's ideas of Sharism circulated in and beyond XinDanWei. After Mao spoke at the internationally renowned electronic arts festival Ars Electronica in Austria, he was invited to talk at several international events in China—from Shanghai's well-known "Get It Louder" event to the aforementioned Chinese Internet Research conference, from Internet blogger conferences to local TEDx events. The XinDanWei cofounders were centrally involved in several of the events at which Mao spoke, serving as co-organizers, hosts, or presenters, and I was therefore able to follow the evolution of the idea of Sharism as it was incorporated into a larger vision of the Chinese Internet and of digital society more broadly.

The panel conversations, talks, debates, and writings that grew out of these events were all key in helping translate the technological promise typically associated with the American free culture movement into the Chinese context. The key argument of Mao's Sharism was that the power of the Internet—from open sharing and peer production to creative and free expression—would transform Chinese society and return it to its "natural state" of being, which, he argued, had been only temporarily interrupted by the limits of "a closed culture" (by which Mao meant both IP and copyright law and Chinese Internet censorship). Fundamentally, Sharism was grounded in the idea that a technology-liberated individual would benefit China not by political action, but by contributing economic growth. As the opening paragraph of Sharism puts it,

SHARISM is a Mind Revolution: The more you give, the more you get. The more you share, the more you are shared. Sharism is an ideology for our Internet Age. It is a philosophy piped through the human and technological networks of Free and Open Source software. It is the motivation behind every piece of User-Generated Content. It is the pledge of Creative Commons, to share, remix and give credit to the latest and greatest of our cultural creations. Sharism is also a mental practice that anyone can try, a daily act that beckons a future of increased social intelligence. It should not go unnoticed that a superabundance of community respect and social capital are being accumulated by precisely those who share.

分享主义是一场心灵革命：你给出的越多，你得到的越多。你分享的越多，你将得到更多人的分享。分享主义，是我们身处的资讯时代的信仰系统。它是透过自由及开放源代码软件的人际与技术网络所传播的哲学。它是每一则用户生成内容 (UGC, User-Generated Content) 背后的动机。它是创作共用 (Creative Commons) 的宗旨，鼓励人们分享、重混，并给予文化创造更多的认同与肯定。分享主义，是一种人人都可以尝试的精神实践，它也是一种日常的行为，召唤了日益精进的集体智慧的到来。我们更不应忽略，乐于分享的人们获得了来自各种社群的敬意，社会资本因此得以不断地累积。[51]

Like the excerpt above, Mao's texts are often vague, but rich in promise; a feeling of revolutionary potential emanates from the pages. What weaves throughout the text are some of the key concepts and values of the free culture movement; the revolutionary potential of social media, the economic benefit of open sharing and remixing of content, attribution, networking, and collaboration. Crucially, if those values were taken up in China, Mao suggests, Chinese people themselves would be redeemed—they would find "respect" in the international community (like he had) and would gain the "social capital" necessary to sustain flexible work. Throughout various texts he produced on the topic, Mao emphasizes that Sharism was "a productive economic force" rooted in the unrestricted flow of information and the economic empowerment of individuals. Sharism was not, he stresses, a political movement comparable to that of socialism or communism:

> The new economic formula is, the more people remix your works, the higher the return. I want to point out that Sharism is not Communism, nor Socialism. As for those die-hard Communists we know, they have often abused people's sharing nature and forced them to give up their rights, and their property. Socialism, that tender Communism, in our experience also lacked respect for these rights. Under these systems, the state owns all property. Under Sharism, you can keep [individual] ownership, if you want. But I like to share. And this is how I choose to spread ideas, and prosperity.[52]

Isaac Mao distinguishes Sharism from communism by grounding his argument for empowerment in Lawrence Lessig's work on remix.[53] According to Lessig, the cultural practice of remix enables those working outside corporations and copyright regimes to develop alternative economic models that reward creativity in appropriating, expanding, and remixing prior work; the producer's creativity accrues social capital that can be leveraged for personal economic gain.[54] In his foreword to Ito's *Freesouls*, Lessig asserts that digital technology further expands "the opportunity for [these] artists to profit from their creativity. They still do what they do for the love of what they do, but they can practice that love more because the market for creative work has expanded."[55] He differentiates this from a time before digital networking when artists were isolated from their audiences and users and therefore struggled financially. Lessig here uncritically promotes as empowering the kind of relational labor in the creative and arts industries that feminist scholar Nancy Baym identifies as deeply exploitative.[56] What Lessig describes is familiar to any scholar of neoliberal techniques of governance: the human is rendered as an asset, her innate abilities and creative capacities framed as valuable when they attract future uptake, remix, and economic investment.[57]

This pitch that self-economization would provide individuals with interventionist capacities and political agency was powerful in the Chinese context, where people had been experimenting with a range of ideas to assert a sense of justice and parity for Chinese people as described earlier. There was an affect of exhilarated excitement around Sharism as endorsements of it and of Mao poured in from Western academic networks such as the Chinese Internet Research community and institutions such as the Berkman Klein Center, Ars Electronica, and Creative Commons. These Western institutions further legitimized the framing of the neoliberal doctrine of self-economization as hopeful and promising for Chinese society. The growing visibility of XinDanWei in these same networks increased this heady feeling. The numerous events I attended alongside XinDanWei, the private debates and conversations that I observed, and the various collaborative projects that emerged from the coworking space further proliferated an affect of agency and hope. I interviewed numerous people who circled through these networks, both from China and abroad. Many of the Chinese I met had returned from studying or working abroad, and the stories they told me echoed those I heard from many of the foreigners: that China was exciting, a place where older projects and aspirations of the Internet could be taken to a next level, experimented with anew. This new China was set in contrast to Europe and the United States, which were described as too slow to adapt or too bureaucratic to change, as caught up in traditions and old value systems, with IP and copyright being the two prominent examples.

At one of these events I met Zafka Zhang. "Creativity is in our DNA," he told me as we sat down over a coffee a couple of weeks after the event. I wanted to hear more. Zhang explained, "we are the first generation in China to realize our dream . . . we do not just want to survive; we want to lead a better life. We believe in individuals, not organizations. Our DNA is for individuality. When I lived in San Francisco, I read Giddens and what he termed risk society. This is where the youths in China are today. They reflect. They depend on their communities. Dream, action, and reflection. These are our three words." In 2006, Zhang had worked for the China-based offices of the online company Second Life, which brought him to the United States for the first time. All of this was "utterly exciting," he explained to me. "We were at the frontier of society." He went on to study public administration at Fudan University in Shanghai and then political studies in London. After his return to China in 2008, he and a couple of friends started a company called "China Youthology." The company was founded on the same ideas he had expressed to me over coffee: that China's young generation was thriving on individuality, and that risk-taking was in their DNA. China Youthology is a platform that connects young Chinese creative designers and artists with international brands. The company positioned itself as a mediator between the big corporations with brands but "no ideas" and youth "who are actively creating new meanings and are shaping the important values of the next generation." According to the company's corporate flyers, its main task was to "dig out the most provocative insights from today's youth with one sole purpose: to give brands accurate inspiration so they can uncover the right direction forward. China Youthology is crowdsourcing for China's future." In other words, what the company sold was access to Chinese youth, rendering them as unique and thus as human capital that offered access to the values and ideals of a whole generation (and thereby to an enormous new market). Companies like Nike, Intel, Coca Cola, and many more invested in China Youthology's promise of access to youth capital.

Many of the people I met during my research with XinDanWei were drawn to such articulations of young Chinese people as creative and as "valuable" to a global market of creative tech production. This excitement must be understood in relation to the CCP's articulations of Chinese citizens' lack of creativity and to Western discourses of modernization and development. As I have elaborated in greater depth in the introduction of this book, both discourses rendered colonial regions and the global south as "other," as lagging "behind" the West in terms of their economic and social development and as lacking the capacity for self-driven innovation and creation. This position legitimized all kinds of Western interventions, from colonial exploitation to foreign aid and development programs. And it was the very narrative of promise and progress that masked continuous resource extraction by the West.[58] It also legitimized the CCP's regulation of the conduct of Chinese people by inducing desires to self-upgrade and self-economize. Entrepreneurial life as articulated through the ideals of an open and free society enabled by Internet technology was powerful in this particular postcolonial context. It cloaked self-economization in a promise of revolutionary action and countercultural heroism, rendered as available and desirable for people in China. The celebratory embrace by the West of people like Isaac Mao further legitimized this view.

How much this countercultural promise mattered became clear when several of these events and projects people had organized in China were censored, with some barely escaping government shutdown. The censorship seemed to prove how important this work was for Chinese society. One such event was the "Designing the Hybrid City" conference that a group of Dutch design researchers had organized in the fall 2010. Liu Yan, Isaac Mao, and one of XinDanWei's members were invited as speakers. The conference was aligned with the concomitant Shanghai Expo and its theme, "Better City, Better Life." The conference organizers argued that Shanghai, and China more broadly, provided a unique cultural vantage point from which to newly explore old questions of mobile life and hybrid living (physical-digital) in the city. But these ideas seemed to challenge something in the CCP, for when I arrived at the conference venue, one of Shanghai's recently established creative industry hubs, official signage for the event was nowhere to be found. When I asked one of the security guards at the venue for directions, he just waved his hand, indicating that he was unwilling or unable to provide any information. I eventually ran into one of the Dutch organizers, who showed me the way to a room off the main venue, on the back side of the building. As the sixty or so participants trickled in, slowly finding their way to the hidden location, the organizer opened the event by explaining the strange location of the conference space: "So, why is there this change of event [space]? We are not quite sure ourselves. But what we do know is that we should keep the event off the radar. This is not an event that is actually happening. So please refrain from using any social media or other public documentation." Conversations between talks and over lunch and dinner returned frequently to the question of why this event had been officially shut down. There were humorous interpretations mixed with concern (mostly expressed by foreigners) for some of the Chinese presenters and the consequences they might face. There was a shared sense that what was discussed at the conference must be important—why else would it have been censored?

The socialist pitch put an affective skim of interventionism and agency atop demands of economic agency. It rendered the redemption of Chinese people as creative and innovative to be within reach. In China, flexible and entrepreneurial work was understood as promising to achieve justice for the Chinese people and even political change, exactly because the West was understood as modeling and thus legitimizing it.

Chuangke (创客) and the Promise of Happiness

One of the approaches that had gained particular traction in this space of prototyping alternatives to the state's construal of citizenship was open source hardware and making. Over time, a splinter group formed—much of it due to the work of two of XinDanWei's early affiliates: David Li and Ricky Ye. Ricky Ye had just returned to his hometown of Shanghai from England, where he had completed a PhD degree in robotics. While in England, he had formed a small transnational working group of Chinese roboticists who had connected through an online forum; the group had worked together from 2005 to 2008 to build an open source robot that would demonstrate both the possibilities of open source and the engineering and innovation capacities of young Chinese men. In 2007, Ye returned to Shanghai and, with some of the collaborators he had met online, turned those early experiments into a new company: DFRobot. DFRobot became one of China's earliest and most well-known open source hardware companies, co-hosting and sponsoring local maker events and receiving invitations to speak abroad and attend international open hardware gatherings. These ideas (and the company) grew out of Ye's first attempts to showcase the first robot built by the group of online collaborators. The robot became the centerpiece of a series of open source hardware and robotics events at XinDanWei. These events were well attended (often a hundred people or more) and sometimes led to impromptu speeches (which would later become more organized, formal speeches) by people interested in open source hardware.

One of these informal speakers was David Li. Li, originally from Taiwan, had moved to Shanghai in the early 2000s after quitting his PhD studies in AI and computer science at USC when his advisor failed to get tenure. Li received numerous job offers from industry, which promised a more sustainable and exciting future than academia. For several years, Li worked in Japan, where he started a company that produced the code and content for the early online and mobile game market. He moved to Shanghai because it was exciting: "China was the place where things were happening," he reflected while I sat down with him over coffee one afternoon at XinDanWei in 2010. "It was very different from my time in the US. All the energy was here [Shanghai]." In the late summer and early fall of 2010, Li began a series of talks at XinDanWei about "Makers: a new movement" and "the power of open source hardware." Just as Ye's robotics competitions had done, Li's talks drew big crowds; over time, XinDanWei became known among people who had been active in open source hardware and making communities in Shanghai and beyond.

It was at a BarCamp that XinDanWei hosted in the fall 2010 when ideas that had been circulating around the potential of open source hardware began solidifying as a particularly promising approach for China's prototype citizen. BarCamps are fairly informal gatherings without a pre-planned program or content. People are invited to participate without formal registration, and participants propose topics of interest. The event, which took place on the weekend, had been promoted through XinDanWei's mailing list, and more than three hundred people had signed up to participate. The day of the event, Chen Xu put up a large white board next to the entrance of the coworking space, and by 11 am, the space was filled with a diverse crowd of people—some of whom I recognized from previous XinDanWei gatherings, along with many new faces. One of the repeated themes listed on the board that day was the question of innovation in China. Ricky Ng-Adam, who had just moved from San Francisco to Shanghai for an intracompany job transfer at Google, offered to give an informal talk on the topic later that day. At 2 pm, about thirty people gathered in one of XinDanwei's smaller conference rooms to listen to Ng-Adam. David Li and I sat in the back row; as Ng-Adam began to talk about how China fundamentally lacked the kind of innovation capacity he saw as thriving at American corporations like Google, I watched David Li practically vibrating in his seat. Ng-Adam began by asking, "So how can we transfer Silicon Valley to China? Here in China, people don't speak up. Is it possible to import this culture into China? Even the culture at Google here in China is different [than at Google in Silicon Valley]. I can't get people to participate and to care about the company here—you know, this 'I am really involved in it and part of it' kinda thing. We need to create that here in China. Here people just work really hard, but they aren't innovative, even though Google has the same 10% off policies here." A Chinese audience member chimed in in agreement: "Well, here in China, there is no world of Silicon Valley, because China is all about execution. You know, there are no coffee houses, no Twitter. When I first came back to China [from overseas], I didn't know where to find these things." Next, an American claimed, "In China you really have to cultivate collaboration. A lot of my team members are anxious to break the silos. You basically have to change [the Chinese] managers." At this moment, Li intervened. What he said that day at the BarCamp crystallized for him a key mission, one that he hoped China's maker movement would follow. China, Li said, did not need to adopt a Silicon Valley approach. On the contrary, it was China's own markets, its existing production—from its factories to the mundane repair culture of China's streets—that constituted and nurtured a China-specific form of innovation.

Although the group didn't settle their disagreements that day, Li's claim that making might be an ideal avenue to redeem China and its people as already creative and innovative (rather than in need of upgrade) stuck with others. After the BarCamp, Ricky Ng-Adam, Liu Yan, David Li, and Min Lin Hsieh (who had attended the BarCamp) began meeting on a regular basis. Conversations evolved around the potential of open source hardware and making, and these conversations gave the coworking space a renewed focus, beginning to shape for it a new, more concrete identity. I attended many of these informal gatherings, and I was in attendance when the group decided to move these ideas into a more tangible realm: Liu Yan, Chen Xu, and aaajiao agreed to rent one of XinDanWei's rooms to Li, Ng-Adams, and Hsieh, creating China's first hackerspace. It was Liu Yan who came up with its name: XinCheJian (new workshop), "to keep with the theme of XinDanWei." Liu Yan explained, "Danwei means workunit, and since we are making stuff, we call it XinCheJian, new workshop or new factory floor."

For the first several months, XinCheJian hosted weekly talks and workshops to introduce the wider network of XinDanWei followers to the ideas of open source hardware, making, Arduino, 3D printing, and so on. Many of these early events took place around one big table set up in XinCheJian's rented room, filled with cables, wires, Arduino boards, breadboards, and more. At this stage, not much making was happening; as the XinCheJian team emphasized, this was a period of "showing what was possible." About six months after XinCheJian opened, the community had grown enough that it needed a larger space. They gave up the small sublet at XinDanWei and moved into a place of their own. The success of XinCheJian provided fuel for other informal maker

FIGURE 2.4. XinCheJian hackerspace, Anhua Road, Shanghai, 2010. Photograph by the author.

and open source hardware experiments that had been carried out in various other cities. Within the timeframe of only six months, eight more hackerspaces opened in Beijing, Shenzhen, Ningbo, Guangzhou, and Hangzhou.

Many of these spaces' founding members were involved in the organization of China's first official maker-related event, the Beijing Maker Carnival, which took place in April 2012. The event was a big deal. It brought together China's fledgling maker and open source hardware community with a group of like-minded hackers and makers from abroad. Many of the foreign makers who attended were eager to hear what China's unique take was on making. Overall, the event (considered a success both in China and abroad) demonstrated that making had much to offer Chinese society—a view that was legitimized by the international attendees. One of them, Mitch Altman, co-founder of the San Francisco hackerspace Noisebridge (which became widely understood as a cornerstone institution in the democratization of tech production via open source hardware and DIY making), was invited to give the keynote.

In his talk, Altman described hackerspaces as the kind of environment that would empower people to "explore what they love, no matter if it makes money." Such an attitude, he argued, was crucial for China's economic future, which he noted had "a couple of challenges." If China "wanted to be part of the modern economy, if it is going to have an economic future," he continued, it "must turn into a creative economy." And in order to build such a creative economy, he argued further, China needs "people who take risks" and "follow their dreams." But there was much hope for China, Altman insisted. Pointing to the people who had organized the event—China's makers and founders of hackerspaces—he suggested that "some people here already know how to do that. They explore what they love." Hackerspaces were crucial, he made clear, to spread this attitude of following your dreams beyond the few people in the room. "The Internet is all fine, it's a great tool, but it's not a real community." Hackerspaces, by contrast, are "supportive communities," he posed, "where people come together and share what they love." "In hackerspaces, magical things can happen, all day long and all year around," he added. Fundamentally, Altman stressed, they enabled people to do "cool things," which enabled them to "make money" off their dreams, and by extension lead China into a hopeful future:

> Right now, China's economy is based on making stuff here and manufacturing it for the West and shipping it back to the West. All of the world economy today is based on creative economy. And if China is going to be part of this economy, people have to be able to take those risks and be encouraged to be creative. If China can do that, if people here can do that, there will be lots of people here who explore and can do what they love . . . This is the first hands-on maker creative-sharing festival in China. It brings the makers here in a worldwide movement of hackers. We all do what we love. And like all makers, we are to enthusiastically learn from one another. We are here to inspire and to be inspired. Creativity is here to launch the country into its future.

In the preceding statement, Altman portrays China as held back by "made in China" and as lacking creativity, ironically echoing the official discourse of the CCP I described earlier. If (and only if) China would—just like the West (and following in the footsteps of the West)—adopt the machineries of the creative industry, so Altman stipulated, it would "launch the country into its future."

Altman's arguments carried weight for the audience. He had become an outspoken critic of *Maker Media*'s acceptance of funding from the US Military Agency DARPA and had ever since refused to attend Maker Faires in the United States. His stipulations that "true" making would only be possible if not attached to machineries of the military-industrial complex gave him an aura of authenticity—a true countercultural spirit, an authentic hacker. Over the years of my research, I have heard Altman give a variation of his 2012 speech at the Maker Carnival (and often literally the same content) at numerous maker-related events in China. All of these events were funded by the Chinese government, and all of them had funded Altman to attend, speak, and

FIGURE 2.5. Mitch Altman speaking at the 2016 Shenzhen Maker Faire (presentation slide reads: "Be a happy worker!"). Photograph by the author.

host educational workshops, teaching Chinese kids and students how to solder. It might seem surprising at first that a countercultural hacker from the United States would garner so much interest and support by the CCP. Altman was one of many other Western maker advocates who were invited to speak over the years in China; adding to the powerful socialist pitch of entrepreneurial life that if people turned themselves into creative producers, the national economy would have a future and people would achieve justice. They cloaked the very demand of the CCP of its citizens to self-upgrade into creative tinkerers and entrepreneurial change agents in a discourse of hope and promise. If Chinese people turned themselves into innovators, they would be able live the modernist promise of the good life[59] and of happiness,[60] so the message of the pitch.

It was precisely because this pitch for an entrepreneurial life was articulated by people like Altman—an American hacker with an aura of countercultural authenticity—that it carried so much weight in China. This pitch appealed to the CCP, because it called upon people to channel their dreams and passions into economic activity. And it appealed to many in China's fledgling open source hardware and maker scene, because it promised a sense of agency and justice and construed Chinese people as able to self-upgrade into passionate makers—it promised happiness. This was brought home, for instance, in the comments by Justin Wang, a founding member of one of Beijing's early makerspaces and one of the key organizers of the 2012 Beijing Maker Carnival, following Altman's speech:

> I am talking about a kind of culture and lifestyle that makes you feel you are happy. How happy you can become with this lifestyle is what we share. This is the identity of the maker. We advocate for thousands and millions in China [to become makers], so that this can grow up in China. We believe

that everyone is a maker . . . think of Steve Jobs who said, "the reason why I developed the Mac (Macintosh) first is I know I can do it and second, I know I can make it perfect." When he was 11 years old, he made his first "kit." When you start thinking about creation and about innovation it is something in your heart. When we think of the lack of innovation in China, that's what you always hear. But we have the greatest opportunity here, we have the factories and the cheapest labor. We are so talented, and we have ideas. All we need to do is make these ideas a reality. And this is what being a maker is all about.[61]

Wang here articulates the dual appeal of making: it was a fun engagement with technology that also cultivated a particular kind of agency for Chinese people. Making would redeem China itself, Wang proposes, as a place of "opportunities, talent, and ideas"—the breeding ground of the "kind of culture and lifestyle that makes you feel happy." Wang's comments hint at a discursive shift in how the self and the nation were being articulated—a shift that Chinese and foreign maker advocates prototyped, in a sense. No longer would China have to be ashamed of its factories and manufacturing sites; instead, it was China that had the necessary concrete "material" from which to prototype alternative futures, and the nation itself would be a laboratory—as Wang put it: "we have the greatest opportunity here, we have the factories and the cheapest labor. We are so talented, and we have ideas. All we need to do is make these ideas a reality. And this is what being a maker is all about."

And this explains making's broad appeal across often opposing sites and viewpoints in China: making became associated with the feeling of self-actualization—it brought into seeming reach a way of being Chinese that was self-defining and happy. This promise of happiness was powerful in China's context of colonial endurance as described earlier, for staking the claim to happiness rejects colonialism and the narrative of lack that it produced. Sara Ahmed describes happiness as inextricably linked to colonialism—in her words, "the civilizing mission of the colonial periods could be described as a happiness mission,"[62] for colonization was justified by constituting the "other" as lacking not only in civility and knowledge but also in happiness. Drawing on critical theorist Homi Bhabha, Ahmed shows that the colonized, by being rendered as "lacking the qualities or attributes necessary for a happy state of existence," are required to mimic the colonizer by approximating their habits. Such mimicry produced a hybrid subject: "almost the same, but not quite, almost the same, but not white . . . almost happy, but not quite."[63]

The self-fashioning as a maker promised to overcome associations of Chineseness with lacking (in civility, in happiness, in autonomy) that had dominated political and international discourse since China's partial colonization. The discourse of wenming—as detailed earlier—demands that China's citizens self-upgrade in order to contribute to implementing a national alternative to Western modernity. It has become a core technique of China's neoliberal governance.[64] If citizens upgrade themselves, the CCP claims, China would be granted legitimacy by and simultaneously be freed from the West's hegemonic reach. As an anti-colonial project to establish sovereignty of the Chinese nation, the CCP ironically mobilized colonial tropes of lack in order to achieve difference and parity (with the West). Making promised exactly that; it offered the articulation of difference (a uniquely Chinese approach to technology innovation, rooted in its manufacturing and labor histories) and it offered parity with the West (if the self was refashioned into a creative maker, who was granted to live the good life). The way making was pitched by many of the internationally renowned figures, who were invited to China's early maker events and who (similar to Altman) created affective links between making, innovation, and entrepreneurial activity—the message was clear: not any form of making, but a particular approach to making, would grant Chinese people the label of happy modern innovators.

The Chinese characters for "maker" that China's open source hardware advocates chose reflect this promise. It was during the planning stages of the Beijing Maker Carnival that its organizers began debating about how to best translate making into Chinese. While it was clear that the English term "maker" should be used to brand the event among makers and hackerspaces abroad, there was not yet a Chinese term appropriate to communicate what the organizers had in mind. The organizers decided that 黑客 heike, the Chinese term for hacker, was out of the question, as it connoted illegal activities; heike translates literally as "black professional" and (like the term "hacker" in English) has negative connotations in vernacular Chinese. After some deliberation, David Li instead proposed the term 创客 chuangke, which is a combination of the characters 创 chuang (to create, to start doing something, to initiate) and 客 ke (professional, expert).

The character 创 chuang is used in 创意 chuangyi (creativity), 创新 chuangxin (innovation), and also 创业 chuangye (start a business). The literal translation of 创客 chuangke can be anything from "creative professional" to "innovative professional" to "entrepreneurial professional." Li argued that chuangke "connotes something positive, like innovation and creativity." After some deliberation, the rest of the organizing committee agreed that chuangke was an ideal term. The choice was clever, for it left open for interpretation what exactly it referred to: it could mean making, entrepreneurship, innovation, or creativity. This term would become crucial in the years to follow, for its connotations of optimism, happiness, and techno-economic agency paved the way for the Chinese government to appropriate making, using it as the groundwork for the series of policies on mass innovation and entrepreneurship (see the book's introduction and chapter 6 for detailed discussions of these policies).

Chuangke articulated a subjectivity that had long been denied Chinese people by both the Chinese government and foreign tech discourse: as Justin Wang articulated it at the Maker Carnival, chuangke stood for "a kind of culture and lifestyle that makes you feel you are happy. How happy you can become with this lifestyle is what we share." According to Wang, "This is the identity of the maker." Chuangke would become a key term that rearticulated the civilization discourse of wenming through notions of techno-optimism, tinkering, playful experimentation, and happiness. By revising Chinese people as chuangke, China's maker advocates positioned China and its citizens as innovative and creative—labels that had previously been attached to white bodies, foreign makers, and Western models of creativity.

As China's maker scene began hosting international gatherings (like the Beijing Maker Carnival) and attracting attention from local governments, international corporations, and Western institutions, earlier ideals of Sharism were fading into the background. Rumor had it that Isaac Mao had moved to Hong Kong after a series of police interrogations, largely related to his BBC interview. In 2013, Liu Yan, Chen Xu, and aaajiao decided to close XinDanWei and turn over the physical space to the event-hosting agency Lohas. They explained to me that the coworking space, which had never been aimed at making a profit, had become financially unsustainable for them. While it was difficult for them to let go of the space, the three co-founders felt they had accomplished much: they (and the groups and networks they had fostered) had prototyped an alternative way of being Chinese, both nationally and abroad; they had modeled how Chineseness can become attached to ideals of happiness, techno-optimism, and the good life—they prototyped how Chinese people can achieve a sense of parity with the West.

The Violence of Technological Promise

In the 1990s, the American free culture movement had depicted technology in revolutionary terms; the world would be saved through technological ingenuity and creativity itself. As Fred Turner shows, the movement's critiques of midcentury bourgeois life ushered in a "new form of that life: the flexible, consciousness-centered work practices of the postindustrial society."[65] They transformed the values of the Cold War research laboratories—their technocratic orientations and their commitments to collaboration, peer production, and social and technological experimentation—into the ideological underpinnings of what became known (not without irony) as the American Internet counterculture. This technological promise—that if people turn themselves into technological tinkerers and countercultural hackers, radical change would be possible[66]—has endured; it is visible in the ways in which Silicon Valley tech start-ups pitch their apps as making the world a better place, in the ways in which companies like Google and even research fields like human-computer interaction frame their work as "doing good," and in the ways in which visions of a global "creative class"—pitched as universal templates[67]—continue to be taken up in smart and creative city initiatives. Indeed, it is the promise of societal and economic transformation via technological experimentation that explains the endorsement of these ideals by governments around the world; policy makers and corporations have embraced makerspaces, incubators, and innovation hubs because they have been pitched as creating difference and "regional advantage"—in other words, they promise other regions to develop Silicon Valleys and a creative class of their own.[68]

We must attend to the exploitation and violence that is occluded when technological promise is displaced into regions long denied the status of creative and innovative. Yearnings for parity and justice, induced by colonial tropes of lack, make people endure the "side effects" of being granted the label of innovator: depoliticization appears to be in people's interests and the demand to channel one's unique abilities and passions into economic future making seems to provide agency rather than proliferate precarity. Precarious conditions of flexible work and the creative industry appear desirable rather than what they are—coercive; as the feminist scholar Angela McRobbie puts it, "the call to unleash inner creativity brings the tantalizing promise of self-reward, thereby almost negating the thread of insecurity."[69] Technological alternatives to the structures of IP and corporate monopoly (in the form of peer production, open sharing, free and open source software and hardware) have fueled rather than intervened in contemporary forms of neoliberal capitalism, exactly because of their countercultural affect.[70] The feeling of interventionist capacity attached to technological alternatives from open source hacking and peer production to coworking and DIY making has masked the depoliticization that occurs when technological tinkering seems to make political action unnecessary.

The socialist pitch cloaks the neoliberal demand of self-economization in a particular kind of affect: fun, playful, happy, anti-bureaucratic, and interventionist (the economization of life appears to be entrepreneurial life). It promises agency within a society governed by an authoritarian regime that has denied political participation to many of its citizens. Displacements of technological promise (from the promise of a global creative class to the pitch of democratized Silicon Valley–like tech innovation) ironically serve a political regime that emphasizes economic over political agency. In China, people were drawn to these ideas of agency and the promise of happiness via technological experimentation, exactly because Chinese citizens had long been denied such subject positions by both the CCP and the West.

3

Inventing Shenzhen

HOW THE COPY BECAME THE PROTOTYPE, OR: HOW CHINA OUT-WESTED THE WEST AND SAVED MODERNITY

The Orient is not simply there. It is made.

—EDWARD SAID, *ORIENTALISM*, P. 5

Orientalism, which earlier articulated a distancing of Asian societies from the Euro-American, now appears in the articulation of differences within a global modernity as Asian societies emerge as dynamic participants in a global capitalism.

—ARIF DIRLIK, *THE POSTCOLONIAL AURA*, P. 108

In 2014, Joi Ito, then-director of the MIT Media Lab, traveled with a group of his students and faculty to Shenzhen. Upon his return, he published a "Shenzhen trip report," describing how the city in the south of China created a "role reversal" via its "manufacturing ecosystem."[1] He claims that many in the West failed to see Shenzhen's innovation for what it actually is:

> What was more impressive to me than the technology were the people [we met] . . . such as the factory boss . . . and the project managers and engineers. They were clearly hard-working, very experienced, trustworthy, and excited about working with [us]. They were willing and able to design and try all kinds of new processes to produce things that have never been manufactured before. Their work ethic and their energy reminded me very

much of what I imagined many of the founding entrepreneurs and engineers in Japan must have been like who built the Japanese manufacturing industry after the war.[2]

Ito here frames the promise of Shenzhen's manufacturing culture in terms of its resemblance to Japan's past. In a TED talk entitled "Want to Innovate? Become a Now-ist" that Ito gave around the same time, he argued that opportunity would be created, if the students trained in Western university laboratories such as the MIT Media Lab could absorb the ethos of hard work and trustworthiness he had discovered in Shenzhen's manufacturing culture; these students would transform themselves into what he called "Now-ist[s]"—people who recognized that "nowadays to innovate you don't need to plan everything, but you need to stay connected, always learning, and always present."[3] He offered the notion of the Now-ist as an alternative to being "a futurist." As he saw it, innovators today needed to be committed to a mode of constant experimentation and continuous learning, to be sharply attuned to the ongoing shifts in the present. Learning from Shenzhen, Ito explained, would provide those prone to become the innovators of tomorrow—such as the students at the MIT Media Lab—with both the tools and (more important) with the ethic and self-discipline he considered necessary for them to self-transform into "now-ist" innovators. For Ito, traveling to Shenzhen was like going back in time and learning from the past, momentarily reliving the foundations of Japan's technological and economic success. This is peculiar. Ito looks to the *past* for the tools that will allow innovators in the *present* (the now) to bring about the techno-utopian *future*.

Ito's TED talk delivered an urgent message: He asserted that the entrenched notions of technological innovation—technology as it is today—must be remade to guarantee that progress remained not only possible but also ethical.[4] Ito's sense of urgency must be read as emerging from a sentiment that had begun to proliferate in various engineering, design, and tech research networks following the 2007–2008 financial crisis. A sense of loss and precarity had begun to haunt this once-privileged class of tech and creative workers whose jobs seemed newly vulnerable because of the proliferation of precarious conditions of work. While precarity was not new in the tech industry,[5] the financial crisis brought this precarity into view, especially in the United States—a sense that the techno-utopian promise of the good life, of control and agency over one's future, were being eroded. This sense of crisis was also deeply intertwined with a realization that the techno-utopia was no utopia: design, engineering, and computing had been complicit in bringing about and strengthening the very structures of capitalist exploitation, control, surveillance, and economization that now threatened the industry and its workers.[6] During this period, as the ethics of computing and design and the promise of modern progress itself were called into question, an increasing number of people from fields as far ranging as engineering, the arts, IT policy, education, architecture, and design turned to Shenzhen. By embedding themselves in the factories, workshops, and supply chains of China, they argued, the technology innovation and design could be recuperated, their virtues and moral underpinnings restored.

This construction of a new China amidst rising uncertainty was powerful; it provided a narrative of technological promise and future making in the very moment when many had begun to see that the tech and creative industries were not exempt from the feminization and precarization that had already devalued the labor and skill in other and "older" industries. From 2014 to 2017, Western media (*Wired* magazine, the *Economist*, *Forbes*, the *MIT Tech Review*, and others) took up the story of Shenzhen as a promising hub of innovation and the "Silicon Valley of Hardware." Within these few years, the story of a "new" China was constructed, one that was promising for the Western designer and engineer precisely because it was understood as "other" than the West. China was viewed as promising because it was seen as a site of fakes, copies, violations of IP regulations and copyright law, and lax rules of law and regulations. China and its people were portrayed as an opportunity to go back in time, to return to a period before the feminization of the tech and creative industries took hold in the West. These visions of a "new" China were articulated by powerful actors, whose voices (often male and often associated with the Global North) were heard by international media, investors, politicians, educators, policy makers, and many more. In the broader (and especially Western-centric) design and tech imagination, China was refashioned as a key site for recuperating technological promise. Stories of Foxconn worker suicides, industrial excess, and low-quality production gave way to stories of Shenzhen as the ultimate Maker City, the "city-size hackerspace," the authentic smart city, the archetype of the circular economy—a new model allowing people to rethink modernity, futures, and progress. China, precisely because it was understood as "other," was now the prototype for adjustment to a radically shifting landscape of global tech production and economic development shaped by debates over the ethics of design, automation, AI, and data science.

In these articulations, China was framed as a temporal other—as promising because it was understood as being stuck in the past. What endured in these stories was the kind of orientalist discourse that—as Edward Said identified—once helped legitimize the European colonial project.[7] What differs in this contemporary discourse of "othering" is that China's "otherness," its so-called backwardness, is portrayed as a space of opportunity; it is celebrated for its difference, viewed with nostalgia, seen as having retained space for technological intervention and economic agency. The colonial trope of othering, of relegating China to the past, to the not-yet-modern, served a particular role: as it became clear to many in the West that the dream of modernity—to "bring about an end of scarcity, an abundance of goods, permanent employment, prosperity and the fulfillment of personal happiness," the dream of "living the good life[8]—was unattainable, and likely always had been, China became the place to dream and see the future "again." Shenzhen became an ideal "laboratory" for prototyping ways to inhabit the postmodern world. It seemed unencumbered by the kind of laws, restrictions, and liberal institutions that governed the West. China's lax regulations and "lack" of intellectual property were suddenly celebrated. They allowed the master to dwell in the illusion of taming the land just a little bit longer, to temporarily bask in the sense of control and relive the narrative of progress that had made Europe, America, and Japan great.

This (re)fashioning of colonial and orientalist tropes of othering in contemporary discourse and practice in open source, computing, design, and engineering does not occur in isolation. Communication scholar Fan Yang observes "fiscal orientalism" that has pervaded American media and political rhetoric since the 2007–2008 financial crisis, i.e., the construction of China as temporal other—a menacing future about to surpass the United States because of the latter's debt to China—in order to make sense of the contradictions of financial capitalism and to restore the imagination of an American sovereign national culture that might "still avoid" the "Chinese future."[9] The orientalist tropes deployed by open source and free culture advocates, designers, and engineers, who had traveled in increasing numbers to Shenzhen in the years following the financial crisis, serve an adjacent function to what Yang calls an "ideological fix"—in this case, to recuperate a Western-centric lineage of technological promise and countercultural idealism in the very moment its ethical and moral underpinnings were being scrutinized. It differs from other forms of techno-orientalism (the imagining of Asia and Asians in hypertechnological terms and as a dystopic future power[10]) for China here is articulated as a place and moment of the past—a past that is framed, however, as more hopeful than the current modern, globalized "present" of tech innovation. China, as I will show, began being portrayed as providing a pathway "backwards" to a time and place not yet tainted by the moral perils of a high-tech, newly dystopian present equated with the Western tech industry. This articulation of China, and of the city of Shenzhen in particular, served the displacement of technological promise—it helped make sense of the ethical dilemmas of the tech industry and to recuperate the promise of open source and the free culture movement to democratize innovation. These narratives grew out of yearnings to "return" to a time when tech innovation was considered hopeful and ethical (often equated—as we will see—with the West). Engagements with Shenzhen were articulated as providing the means to revive the technologist, engineer, and designer as an ethical, cosmopolitan, and global citizen (by returning to "hands-on" technology production, deeply engaged with the processes of industrial production and manufacturing, and committed to producing "actual" change rather than abstract ideas) in the very moment that digital technology, including the ideals of open source and peer production, was increasingly contested.

This chapter unpacks in historical detail the making of this "discursive construct" of a "new" (and still "other") China. The postcolonial studies scholar Arturo Escobar uses the term "discursive construct" to describe how the Third World was articulated ("invented") as the West's other, as marked by "poverty" and in "need" of economic development via Western intervention. Escobar traces the emergence of this framing after World War II when the West was strategizing its continuous domination and control in a decolonial/postcolonial world order. This construct of the "Third World" as a problem space, as saturated by poverty, as "unhappy," and "as having needs" took place according to Western categories, an "objectivist and empiricist stance that dictates that the Third World and its people exist 'out there,' to be known through theories and intervened upon from the outside."[11] Technology and science were framed as neutral, desirable, and universally applicable to solve the problems of the Third World. They were portrayed as "bringing light" and as a moral force to produce "happy results" and "awaken" the Third World from its lethargic past.[12]

The "invention of Shenzhen"[13] as an ideal laboratory to experiment with alternative technological futures, this chapter shows, retains some of the same colonial tropes of othering and Euro-American-centric notions of modern progress that were also at the heart of the twentieth-century development project; they portray China through notions of "going back" in time, as a place that was other and less modern than the West, through tropes of adventure and exploration. They presented Shenzhen as a unique laboratory that operated at the scale of a region where global supply chains operated on a daily basis, allowing Western designers, tech firms, and engineers to reinvent themselves and their industries.

What differs is that in this contemporary tech discourse, "otherness" serves the argumentative function of "uniqueness" and "difference." Orientalism and colonialism endure, with otherness portrayed as productive and promising: "otherness" is "difference" fueling the machineries of capital investment that gravitate toward those regions, cities, zones, and other spatially bounded entities that differentiate themselves and are made attractive for future economic growth and capital gain.[14] Postcolonial studies scholar Arif Dirlik described this as follows: "Orientalism, which earlier articulated a distancing of Asian societies from the Euro-American, now appears in the articulation of differences within a global modernity as Asian societies emerge as dynamic participants in a global capitalism." But just like the construction of the Third World legitimized intervention, so did the construction of China as innovative "other" legitimize a range of interventions from foreign venture capital investment to a range of neoliberal government initiatives (a thread I pick up in chapters 4–6).

The seeming celebratory pitch of China as newly innovative and as promising masked the displacements of people deemed incapable of innovating and the racialized and gendered exploitation these interventions brought with them. Orientalist tropes of othering endure via displacements of technological promise. They live on in the refashioning of China as newly innovative, despite a coming to terms with the broken promises of modernization.

But how had Shenzhen's image of copycat been reworked from something that China should overcome into something worth recuperating? How did the copy become an opportunity? Key to this discursive shift, I show, was the rescripting of shanzhai (山寨), China's partially illicit and experimental production culture that had long been decried, both nationally and abroad, as low quality, as fake, as nothing more than economic pursuit, and as emblematizing China's inability to innovate. Shanzhai was portrayed as uniquely Chinese, with Shenzhen essentialized as its ultimate hotbed. It was celebrated as having developed independently from the West, although in fact, China's history of shanzhai production is deeply intertwined with foreign direct investment and outsourcing. Shanzhai emerged in the Western tech imagination as an alternative to individualistic notions of authorship, ownership, and empowerment, and the copy became rearticulated as an alternative to Western-centric notions of design and innovation.

Importantly, this discursive construct of a new China was not without its contestations; the disbelief in modern ideals of progress and creativity coexists with desires to retain and hold on to these ideals. Shanzhai was seen as disassociating innovation and creativity from Western notions of the individualized inventor, and it was therefore considered by some as threatening to the masculinized and individualized entrepreneurship culture of Silicon Valley. This chapter examines the contradictory commitments and desires that came together in the invention of Shenzhen. Like orientalism itself, the "invention of Shenzhen" I examine in this chapter is not a single story of Western domination.[15] The story of a new China was co-produced through desires to recuperate the virtuous promises of technological intervention (desires that were particularly pronounced in the American tech industry) and yearnings for parity and justice for the Chinese people that had driven a range of technological experiments in China as I described them in chapter 2. To illustrate, I draw from three sets of data: first, observations I conducted over the time span of six years (2012–2018) at a range of events and gatherings in Shenzhen (including among others: Maker Faires, Start-up Weekends, entrepreneur meetings, hackathons, design and art exhibitions, the Bi-City Biennale of Urbanism\Architecture in Shenzhen—UABB, panel discussions and debates at a range of small, informal workshops); second, discourse analysis of the writings and talks by people who had come to be associated with both the global maker movement *and* the city of Shenzhen (this included both Chinese and foreigners); and third, conversations and interviews I conducted with China's maker and open technology advocates who had turned toward shanzhai production and Shenzhen around 2008–2010, and with the designers, engineers, self-identified hackers and artists, politicians, and government officials who traveled to Shenzhen from abroad (mostly the West) between 2008 and 2018. This chapter lays important groundwork for the chapters to follow (chapters 4–6)—the discursive construct of a "new" China (via the invention of Shenzhen) I examine here paved the way for a range of urban renewal projects and foreign venture investments in China that were couched in a narrative of promise, happiness, and future making, occluding displacements of people, proliferating precarity, and labor exploitation.

The Tech Industry's Broken World Thinking

Following the financial crisis, many in the American tech industry began reckoning with the fact that many jobs in the IT and creative industries had been feminized (seen as menial jobs that were deskilled and low-paid). Many of those who retained their sought-after jobs worked on precarious short-term contracts with no health insurance or retirement benefits.[16] Lucrative jobs in the IT and creative industries were scarce and available mostly to highly educated people[17]—a far cry from the vision of a global creative class, open to anyone.[18] (This was a double blow to the non-elite class, which was already disproportionately affected by the financial crisis.[19]) The kind of precarity that had long afflicted people of color, including vast populations in the Global South; the rural and urban poor and working class in otherwise wealthy postindustrial societies; and queer and transgender minorities came to be increasingly felt by non-marginalized people in the United States, who were realizing that while "precarity is the condition of our time, . . . we [had] imagine[d] precarity to be an exception to how the world works."[20]

Technological progress had, in appropriately postmodern fashion, become unreliable. What literary scholar Lauren Berlant calls "cruel optimism"[21]— the act of holding on to, remaining attached to the dream of a better future that is almost certainly unattainable—was the constant state of many, and the unattainability of that optimistic future was becoming glaringly obvious. In the creative industries, tech sectors, and in the fields of design and computing,[22] internal debates began, examining how the tech industry's norms, values, and methods had enabled (rather than freed us from) the ills of modernization. Approaches such as design thinking and human-centered design were sharply criticized for their commodification of politically motivated design.[23] Makers of technologies and infrastructures (designers, engineers, computer scientists, architects) began to suspect that they were no longer in control of the systems they made.[24] Digital technology, embedded in and governing capitalist structures of economies of scale, global supply chains, and large-scale infrastructures, began to be seen as inevitable, as rendering futile the kind of direct intervention and resistance that once had seemed possible.[25] The rise of machine learning and the particular flavor of Artificial Intelligence it enabled has proliferated a sense that established forms of resistance and intervention from social movements to opening up the "black box" of technology and participatory approaches to design are ill-suited in the age of free labor, digital labor, algorithmic inequality, and surveillance capitalism.[26]

What had proliferated was an expanding sense that the promise of modernity itself was broken to begin with. Information scholar Steve Jackson expresses these sentiments this way: "Stories and orders of modernity (or whatever else we choose to call the past two-hundred-odd years of Euro-centered human history) are in process of becoming apart, perhaps to be replaced by new and better stories and orders, but perhaps not . . . the 'modern infrastructural ideal' is increasingly under threat, as cracks (sometimes literal ones) show up in our bridges, our highways, our airports, and the nets of our social welfare system."[27] This recognition of modernity coming apart demands, Jackson proposes, "broken world thinking,"[28] which advances a "more hopeful approach: a deep wonder and appreciation for the ongoing activities by which stability (such as it is) is maintained." For Jackson, this includes sites of technology practice typically not celebrated as sites of innovation: repair, maintenance, and infrastructural care.

Amidst such calls for broken world thinking, regions that had previously been seen as the "technology periphery"[29] became the sites of what the *Economist* called in 2010 a "redistribution of hope."[30] This was not only visible in the interest people took in Shenzhen but also in other regions once relegated to the "tech periphery":[31] in 2016, Facebook CEO Mark Zuckerberg traveled to Nigeria and Kenya, which he famously described as examples of "Africa's emerging IT ecosystem." Zuckerberg, here, echoed a flurry of media reports about the region that have begun celebrating Kenya as the "Silicon Savannah" for its advances in digital finance, tech incubators, and local IT innovations such as BRCK and the Ushahidi crowdsourcing platform, among others. Regions that had long been viewed as lagging behind were suddenly positioned as the sites where alternative futures were now graspable.[32] The story of the invention of China has to be understood as emerging from within the context of the tech industry's broken world thinking. As the philosopher Anna Greenspan puts it, "what once seemed to define America, was today easier to find in China."[33]

The turn toward Shenzhen to identify potential alternatives to the perils of technology's complicity in neoliberal capitalism is prefigured by various historical attempts to move beyond the West to remake it.[34] The rise of the so-called four Asian tigers (Hong Kong, Singapore, South Korea, and Taiwan), for instance, sparked numerous development theories in the West, which were interested in uncovering alternatives to the neoliberal doctrine, including most prominently in the 1980s, the efforts by Chalmers Johnson and Robert Wade. Their goal was to learn from the way industrialization had taken place in Asian market economies, which they characterized as developmentalist economies because their development had been guided by the hand of the state rather than emerging from a free market.[35] Johnson's notion of the developmentalist state was an attempt to move away from the "typical Western binary models" (such as Weber's distinction between a "market economy" [Verkehrwirtschaft] and a "planned economy" [Planwirtschaft]), by demonstrating that all states intervene in their economies. He described the 1960s formation of Japan's "developmentalist state" as a "complex process of private-public interaction" that attracted foreign investment, deliberately economically advanced strategic industries, and imported the necessary technology.[36] His particular focus was on the Japanese ministry of international trade and industry (MITI) and how it devised an industrial policy aimed at enhancing the nation's international competitiveness in economic development. "The state took on regulatory function in the interest of maintaining competition, consumer protection, and so forth."[37] The so-called Japan Miracle of 1962, Johnson insisted, had nothing to do with the "cultural essence" of the Japanese but was due to the emergence of the developmentalist state—which, he speculated, could also constitute a model of economic development for the West.

The question of whether models of economic development are transferable from one region to another continues to be hotly debated. China is seen as an edge case in these debates, because of its partial colonization, its communist past, and its authoritarian government. I do not aim to offer yet another answer to this question, for I think it is the wrong question. What interests me is how certain regions and spaces become seen as models or prototypes, and how this process unfolds in our particular context of broken world thinking. Following the financial crisis in 2007–2008, Shenzhen has become such a model, touted as a laboratory of future making for other "developing" economies in Southeast Asia, South America, and Africa.[38]

Shenzhen has drawn the attention of policy makers, development organizations, and investors, and of scholars, activists, and writers interested in the question of whether and how resistance is possible in the context of capital's seemingly endless reach.[39] Scholarly interest in Shenzhen has been shaped by a growing suspicion of utopianism, of the notion that there is any ultimate, universal alternative to the status quo.[40] It is motivated by the recognition that the opposite, a turn toward "petits recits" (Lyotard's term for the postmodernist challenge to meta-narratives of change and progress, and the idea that local and small-scale communes and cooperatives would be sites of resistance) has been plagued by elitism and exclusions—an utopia in the small that appears to simply expand a temporary feeling of "do good" for a small group of people.[41]

Here, too, Shenzhen's appeal stems precisely from its image of a laboratory that operates on a mass scale where alternatives to capitalist structures can be prototyped. Speculations over the forms of resistance that are possible in the age of finance capitalism often turn toward scale. Some have argued that resistance movements would have to adapt to the kind of scale-making practices of capitalism itself in order to have any impact.[42] The turn toward Shenzhen has to be understood as deeply intertwined with these multi-sited yearnings for alternative worlds—visions of the future that remain caught up in the ideology of technological progress and scale. Despite widespread yearnings for alternatives to capitalism, it remains difficult to let go of the ideal of progress all together.

SEZ and Shanzhai (山寨)

Shenzhen's appeal has much to do with the region's history as a SEZ (Special Economic Zone) *and* its intertwined history of grey market production (referred to as *shanzhai* 山寨 in Chinese). When Deng Xiaoping declared Shenzhen a SEZ, it became a "regulated space of political economic experimentation"[43]—a laboratory to test how far China could open up to the globalized markets of capitalism without the communist party state losing credibility (see the introduction chapter for more details). In the SEZs, ways of living and working that had been punishable under socialism (often with one's life) were not only permitted but demanded. Activities like private ownership and entrepreneurial activity were encouraged in order to attract FDI (foreign direct investment), as were incentives like providing foreign companies holidays from income or sales taxes, prohibiting labor unions, being lax in enforcement of (or absence of) labor and environmental laws, allowing access to cheap domestic labor and cheap physical space, offering exemption from import/export duties, and so on.

At first, the SEZ was a "front, while the purity of the state was maintained,"[44] but the capitalist experiments in the SEZ eventually expanded to the rest of the nation. Political scientist Mary Gallagher shows the role FDI played in this process. In the 1980s, FDI was introduced as a dual system that worked alongside SOE (state-operated enterprises), creating a system of competition between the SOEs and FIEs (foreign-invested enterprises), and between SEZs and other regions where privatization and investment were not yet practiced. This "dual-track" system was crucial in the transition from a socialist to a capitalist market system as it instilled desires for economic and social upgrading that enabled what Gallagher calls "contagious capitalism." The competition between the experiments with capitalist market reform in the SEZs and the socialist models and organizations that persisted everywhere else in China led to a race to implement more flexible labor policies, to create a mobile workforce, and to grant autonomy to enterprise. Both individuals and firms found that—as Gallagher puts it—"to stand by and hold fast to 'socialist enterprise' would have meant losing out on the chance to gain not only capital and technology but also prestige from association with the international economy. Economic reform is pushed ahead dynamically by such competition, while resistance is reduced."[45]

This governance through affect (rooted in inducing desires for self-transformation via exceptions and competition) is not unique to China,[46] although China's SEZs are often thought to be particularly successful.[47] Both the World Bank (WB) and the United Nations (UN) have promoted economic zones as a tool for nation-building and free trade in developing countries in order to enable them to enter the global marketplace.[48] The SEZ is understood variously as a "neocolonial spatial formation,"[49] a "differential form of governance,"[50] a technique of "extrastatecraft,"[51] the "quintessential apparatus of the neoliberal state,"[52] a "political instrument" of economic development that masks political intentions,[53] and as "built for speed."[54] But it is more than a simple economic strategy; the SEZ is a technopolitical instrument of affect that mobilizes "imagination, hope, aspiration, and desire." It is thus an ideal political tool, for it affectively masks labor exploitation and environmental destruction by inducing desires for prosperity and the fulfillment of personal happiness.[55] As anthropologist Jamie Cross argues, SEZs are thus "deeply affective spaces in which the future is felt, encountered, and inhabited."[56]

It is the SEZ's affect of experimentation—its associations with prototyping at scale and at speed and its capacity to mobilize desires and hopes—that explains Shenzhen's widespread appeal. The architect Keller Easterling, for instance, sees the SEZ as constituting an urban space that operates like software, a "spatial software" that provides "entrepreneurs" the ability to "hack" capitalist structures, despite their enormous scale.[57] I show in this chapter how Shenzhen became saturated with a feeling of experimentation—the tool of both the state and the "hacker." What contributed centrally to Shenzhen's appeal was the region's illicit electronics production that had flourished in the experimental zone: *shanzhai*. While shanzhai is often understood as a shorthand for "the broader cultural phenomenon of Chinese copycat," it is strongly associated with the electronics manufacturing industry in Shenzhen, where it emerged out of the region's "local, regional and global flows of people, finance, and technology."[58] For example, Shenzhen factories had long been China's main producers of shanzhaiji (山寨机, literally "shanzhai machine"), i.e., a wide variety of mobile phones, which began flooding both national and international markets in 2004. Shanzhaiji manufacturing includes small-batch production of niche phones (for example, dual sim card phones for migrant workers and phones in unique shapes and sizes, some designed for low-income populations and some for elite customers). It also included white-label devices—unbranded devices, fully designed and manufactured in Shenzhen, which companies can simply brand with their logo. These phones were designed and made not only for "emerging" consumer markets in the Global South but also for "established" markets in Europe and the United States. The most controversial version of shanzhaiji (which constitutes one of the industry's smallest market shares) was its copycat iPhones and iPads. These were released as low-budget versions of the Apple products, and they often were sold (sometimes with added features that targeted niche markets in China) even before the official Apple products came to market in the United States. Shanzhaiji became an object of interest as it made shanzhai easier to grasp, for it made visible in the flurry of phones the sheer variety and speed of Shenzhen's production.

Although shanzhai has been written about by a wide variety of scholars and journalists, there is little agreement about how the term was first applied to describe the phenomenon of informal electronics production in Shenzhen. Shanzhai translates into English as mountain village, bandit fortress, or mountain stronghold, understood as referring to "various outlawed but communal forms of self-preservation and self-protection that strove for local autonomy during difficult times throughout Chinese history."[59] The term began to pervade vernacular Chinese around 2007 or 2008. It made the news in China in 2008 when an online media project proclaimed its intention to host a "shanzhai" version of the official CCTV gala, a glamorous Chinese new year celebration, aired annually on Chinese television and featuring prominent stars in music, movies, fashion, the arts, and more. The shanzhai CCTV gala on the Internet instead featured everyday citizens with mundane accomplishments that were rarely publicly celebrated. When it became apparent that the organizer behind the shanzhai gala had commercial interests, the central government intervened and took the shanzhai online version down, but China's Internet sphere came to the organizer's rescue, promoting the project as antiestablishment in its recognition of the heroism of the everyday man.[60] Both national and international media picked up the story, and in 2008, shanzhai was the most often searched word on Google.cn[61] and had become associated with ideals of grassroots innovation and resistance.

The story of the shanzhai gala reflects the broader ambivalence I found many Chinese people have toward shanzhai. While shanzhaiji's rapid spread to markets around the globe fueled nationalistic sentiments about the expansion of China's economic and political power, it was associated with infringement of international copyright and IP regulations, and many Chinese feared that their country would not be seen as a legitimate global innovator. Shanzhai is seen as contradictory—both as an example of China's ruthless copycatting and as a violation of IP and a unique form of grassroots ingenuity, what anthropologist Lyn Jeffery calls a "clever survival tactic." Shanzhai encodes an anti-authoritarianism that is "not unlike that of Robin Hood," Jeffery argues—a kind of ethical criminality that steals from the establishment for the benefit of those excluded from systems of privilege.[62] Shanzhai is an example of "nongovernable subjects," media scholar Ned Rossiter suggests in a similar vein—subjects who are "external to the logistical media of coordination, capture, and control."[63] In other words, shanzhai became understood as demonstrating that grassroots resistance working in the gaps of global capitalism, escaping (as Rossiter put it) full capture and control, was possible. This sense of interventionist capacity attached to shanzhai was key to the crafting of a "new" China in the Western tech imagination. Shanzhai appeared simultaneously to confront the broken promises of modern progress and to restore them, to confirm China's inability to innovate and its ability to produce a more authentic hacker culture.

In this chapter I unpack how shanzhai (as an idealized form of grassroots experimentation that operated across a range of scales, from local to vast) and Shenzhen (as its idealized place of origin) together became a crucial site to work out a series of controversies that were unfolding both in the fledgling open source hardware and maker scene and in the tech and creative industries broadly. I continue by unpacking how China's makers turned toward shanzhai to shift China's reputation from backwardness to innovation, hoping to change China's place in global tech networks. I then turn to some of the foreign actors traveling to Shenzhen and how shanzhai became central in their efforts to reposition their own work amidst heightened debates over the ethics of open source, design, and computing.

Hacking with Chinese Characteristics

> We are in an era where we are starting to re-examine China, this developing country, the copy, and what they [the Chinese] make. We are in an era where we are starting to look into the patent and copyright system. Are they just? Are they ethical?
>
> —DAVID LI, FIELDNOTES BY THE AUTHOR, 2013

China's makers and open source advocates had begun taking an interest in shanzhai as early as 2008, when the debates about what shanzhai production meant for China's global image were all but settled. At that time, shanzhai predominantly connoted embarrassment, a loss of face caused by "Chinese people's apparent sole interest in money" and their "disdain for the general disregard of copyright law."[64] To China's makers, however, shanzhai had something else to offer. As many people active in international maker and open source hardware networks repeatedly demanded that China's makers articulate their uniquely Chinese approach to making (see chapter 2), China's makers saw in shanzhai an opportunity to articulate exactly such difference. One of the early champions of shanzhai was Eric Pan, cofounder and CEO of the Shenzhen-based open source hardware company Seeed Studio. Pan began positioning Seeed Studio as emerging out of a commitment to "combine [the] micro manufacturing efficiency of shanzhai with open innovation and open source," a partnership that, he argued, would "change how China will innovate."

Pan's message found traction in China's fledgling maker community. In December 2010, the Shanghai-based hackerspace XinCheJian invited him to present at a new event series it was hosting under the banner of XinShanzhai 新山寨 (new shanzhai). The XinShanzhai event was the brainchild of Liu Yan, co-founder of the coworking space XinDanWei, and David Li, co-founder of XinCheJian (see chapter 2 for details about both spaces and their roles in articulating China as a site of innovation and creativity). As the term "new shanzhai" implies, the event aimed to give shanzhai another layer of meaning, one that was detached from its negative associations with Chinese backwardness. Pan's own work was ideal for this purpose. Pan articulated shanzhai as having deep roots in a Chinese history of ingenuity and technoscientific experimentation; about halfway into his talk at the—by then already prominent—Shanghai hackerspace XinCheJian, he showed an image of an old painting depicting a Chinese man surrounded by mathematic instruments, sketches, and a vessel emitting steam. "This is Mozi," he explained. "I like to explain shanzhai by referring to Mozi. He was a Chinese philosopher who lived thousands of years ago, around the time of Confucius. Mozi was not only a thinker, he was also a maker. He was a very independent thinker and maker. He wrote about his observations of the world, but he also conducted experiments and built things. He was not wealthy, but he was extremely skilled in creating mechanical things and devices. He was resourceful and self-reliant. This is exactly what shanzhai is."

Pan had struck a nerve. The group that had formed around Liu Yan and David Li had worked hard over the previous years to dispel the notion (which was common both in China and abroad) that Chinese were inherently uncreative, mindlessly copying other people's ideas. Pan's talk sparked a two-hour debate over questions of copy, creativity, and Chinese manufacturing, which was livestreamed on the microblog platform Sina Weibo with the hashtag #新山寨 (new shanzhai). Adding "new" to the term shanzhai sent an important message. It indicated that shanzhai and China's maker movement were not copies but iterations—updates to the visions of a global maker movement. By articulating shanzhai as rooted in Chinese history, values, and practices (the reference to Chinese philosophy is significant here), Pan framed shanzhai as something to be proud of, as the thing that made China's hacker culture unique. Pan unpacked this for me in an interview I conducted with him in 2012: "China's

正传

墨子是历史上唯一一个农民出身的哲学家，墨家创始人。 在先秦时期创立了以几何学、物理学、光学为突出成就的一整套科学理论，号称是位可以囊括所有古代诺贝尔奖的人。

二千多年前墨家便已有对光学（光沿直线前进，并讨论了平面镜、凹面镜、球面镜等，尤以小孔成像出名）、数学（已科学地论述了圆的定义）、力学（提出了力和重量的关系）等自然科学的探讨。

并且，把这些理论大量变成了现实并投入实际运用，如大量的守城兵器。

FIGURE 3.1. Eric Pan's presentation slide depicting Mozi. From Eric Pan.

history of copycat is nothing to be ashamed of. On the contrary, copying means learning. It means you are redoing it, but in your own way. It's like learning a language. You have to write the words and sentences over and over, copying from the teacher. In the process, you learn the basic skills. After you learn the basics, you can create sentences and build grammar, and at one point you can write your own article. You can have your own style. It's a very natural process. It's nothing to be ashamed of or blamed for."

This rescripting of shanzhai aimed to disassociate "Chineseness" from colonial tropes of backwardness and from being seen as "just" copy or fake and instead connect it to both innovation and morality (because it was an alternative to rapacious Western capitalism)—shanzhai was authentic hacking. David Li elaborated this as follows: "What China offers is an alternative model of doing entrepreneurship, an alternative way of doing open source. Rather than this exploitative approach of venture capital and emotional investment that companies like Google prescribe, and this form of creationist capitalism you got with the Web, here [in China] it is the streets, the street markets that innovate. Here, people make money from working together on selling products, rather than selling ideas to VCs [Venture Capitalists]." Li here articulates

shanzhai as constituting a more ethical version of open source and tech entrepreneurship than the Silicon Valley approach, which was becoming increasingly understood as complicit with the machineries of finance investment,
labor exploitation, and proliferating precarity.

In the numerous talks and interviews Li was soon invited to give, he began
referring to shanzhai as "hacking out of necessity" and as "open source hardware in practice." He called shanzhai the "long lost twin of open source"—a
form of open source that was mundane, pragmatic, driven by economic necessity and survival. Shanzhai thus was, he argued, "different from the West where
open source hacking only exists in theory. Here, the actual maker in the factory is involved—the workers, the repair guy on the street." While Pan had
articulated shanzhai by referencing the Chinese philosopher Mozi, Li drew
connections to the Chinese folklore stories of "Shuihuzhuan" (水浒传 literally translated as "Water Margins"), which some speculated had motivated
the term shanzhai (mountain stronghold) in the first place. The Shuihuzhuan
is the story of 108 rebels who were operating independently in the mountains
of Southern China, far from the emperor. Li called Shuihuzhuan the "Book
of Shanzhai," as he explained it to me, "the 'Book of Shanzhai' is about 108
heroes who break regulations, but they are doing it to *right the wrong*. They
are prosecuted for it when they congregate among themselves in the mountain
fortress. But then, in Chinese tradition, there is always hope that *one day their
actions will be recognized*."[65]

For Li, shanzhai's appeal lay in its underlying narrative of the hope for justice—
the "hope that one day their actions will be recognized." What was at stake
for Li was the redemption of Chinese society and culture writ large. He and I
often spoke about his upbringing in Taiwan, the political and economic shifts
he had experienced there as a young, politically active man, and how his move
to the United States after high school had begun to temper his idealistic take
on the world. He began looking at Taiwan and his own aspirations from the
fresh vantage point of a transnational, and he suddenly saw Taiwan's aspirations for independence as undercut by the vast and elitist business networks
in manufacturing and electronics production that connected Taiwan, China,
and the United States. Taiwan's insistence on *political* independence despite
its *economic* co-dependence suddenly seemed to him like hypocrisy. When he
moved to Shanghai via Japan, in early 2000, he sensed that Chinese culture
was being redeemed; he saw this as an expression of "true" independence (in
other words, independence from the West), and this felt intensely exhilarating
to him: "China—well, Shanghai—was the place where things were happening,"
he told me. "In Shanghai, there was all this energy and buzz. People were
trying to put crazy things out there. Shanghai was very experimental. People
would march on Maoming Lu in Cultural Revolution costumes. It was fun to

be here." He paused, then added, "It was very different from my time in the US. All the energy was here [Shanghai]."

When David Li co-founded China's first hackerspace in 2010, he saw it as an opportunity to implement these ideals in practice. He began referring to XinCheJian as a "hackerspace with Chinese characteristics," playing on the popular phrase "socialism with Chinese characteristics," a post-Mao rhetorical formulation used by the CCP to align modernization strategies (such as foreign investment or special economic zones described earlier)[66] with the promotion of national autonomy and independence. In other words, the phrase captures China's intention to modernize without being Westernized[67]—to remain in alignment with its own "core cultural values and traditions."[68] For example, the Party's revival of Confucianism, one such cultural tradition, posited China's past not as holding back its future of capitalist modernization but as *serving* that future. Arif Dirlik describes these processes as being "motivated by a desire to make claims to alternative modernities."[69] It is also a form of self-orientalizing, he argues, because "Asian tradition" is used to legitimize the underlying values of Western modernity tied to capitalist expansion. In the long run, Dirlik warns, this project perpetuates existing forms of power by participating in "nationalist thought" that accepts an essentialized distinction between "West" and "East"—the same objectifying binary constructed in the post-Enlightenment age of Western science. The formulation of "hacking with Chinese characteristics" hints at how the dream to achieve justice for Chinese people has become deeply intertwined with the CCP's nationalist project to redeem the nation on a global stage (and assert party leadership as necessary to do so). It has become increasingly difficult, in other words, to differentiate between the people's yearnings for parity and justice *and* the interests of the party state. Pan and Li—alongside many other maker advocates I met over the course of my research in China—continuously struggled to respond to the (largely Western) accusation that China's makerspaces were just cheap copies of the West's and that they were more interested in making money than in "creative play." It was through such enduring tropes of othering that the CCP's own nationalist interests in articulating a uniquely Chinese approach to development and the hopes of China's maker advocates were driven closer together.

The majority of Eric Pan's early efforts with Seeed Studio focused on distancing the company from Shenzhen's (and China's) association with the fake, the copy, and with low-quality production. Much of this was motivated by incidents like the following—in Pan's words: "When I came to the US for the first time in 2011 to attend the Open Hardware Summit in New York, people there knew us [Seeed Studio] and liked our products, but nobody wanted to believe that we are a Chinese company. Nobody had thought that cool and innovative products could come out of China." When Pan returned to China that year, he began working hard on creating a public image for Seeed Studio

FIGURE 3.2. "Innovate with China" label by Seeed Studio. Photograph by the author.

that aligned it with the values of Western open source hardware—attribution, open sharing, experimental play, and unhinged creativity—and that positioned it as a producer of quality design. He changed the company's product labels from "made in China" to "innovate with China."

Seeed Studio, in partnership with the Shenzhen makerspace Chaihuo, applied for an official license from the company Maker Media in San Francisco to host Maker Faires in Shenzhen and worked for months to receive permission from the Shenzhen government to host them. Seeed Studio also funded makers and members of well-known open source hardware companies and organizations (ranging from Sparkfun and Arduino to the NYU ITP program, the MIT Media Lab, and the Chicago IIT Institute of Design) to visit Shenzhen and speak at various events. Over time, the Shenzhen Maker Faires drew larger and larger audiences, from both China and abroad. I attended all of them and met many of the foreign visitors who came. While many of these visitors celebrated Shenzhen for its scale and speed of production, they complained that the Shenzhen Maker Faire just wasn't the same as the Maker Faire in the Bay Area. They did not like that the Shenzhen version showcased actual "businesses" rather than the kind of creative fun and experimental play that Maker Media had attached to making—"the Chinese," I was often told, need to learn how to focus on being creative first and on making money second. It is hard to overlook the irony of such claims—Chinese approaches to making were stipulated as copycats, while simultaneously called upon to model themselves after (to copy) American maker values and approaches. It becomes painstakingly clear how exclusive this (predominantly) Western vision of making was—the type of playful tinkering and experimentation advocated here excluded most people struggling to make a basic income.

In 2014, Dale Dougherty, the founder of Maker Media, was invited to give the keynote at China's first featured Maker Faire in Shenzhen. Dougherty opened by showing a video from the latest Maker Faire he had hosted in San Francisco. The flurry of images seemed to demonstrate the unhinged creativity of the American maker movement; it depicted people driving in self-made cars shaped like muffins, an iron dragon that spewed fire operated by two people on a bicycle, a young female maker who hacked her own scientific instruments. In his talk, Dougherty explained that all of these examples were what making was fundamentally about:

> Making has this quality that makes it an adventure. It's something fun to do. We don't know how it will turn out and if it will be successful. So, making as an adventure is what attracts so many people. I don't know if I can do this, so let me try. [. . .] In 2006, we organized the first Maker Faire. I didn't know what to expect, but I certainly did not expect to come to a Maker Faire here in Shenzhen, when I launched that. We wanted to share the crazy things that people do, like these electric muffins. You really have to come and experience it, it's much bigger and bolder than you even can imagine. . . . it's a wonderful and incredible event that makes you think this is some of the best side of human beings.[70]

The underlying message here was clear for the largely Chinese audience in the room; that in order to be considered creative by the West, Chinese people must remake themselves in this image: the cheerful, daring, playful, fun, happy maker and risk taker, freed from economic interests and concerns.

Seeed Studio was considered by many a success in this regard—in American networks of open source hardware and making it had become known as an advocate of "authentic" open source—as valuing attribution and advancing people's dreams and creativity. It was celebrated as a Chinese company that demo-ed what Chinese innovation could look like—open source "with Chinese characteristics." When Eric Pan was named one of China's 30 under 30 entrepreneurs by Forbes China in 2013, he was posited as a role model for others in China—a young Chinese man who had redeemed China and its people.

Such (Western-endorsed) stories of success (and of redemption of the Chinese people as creative dreamers) were ideal for the CCP's own purposes (see introduction and chapter 6). By re-articulating shanzhai in terms of notions of (East-West) difference and uniqueness, China's makers had responded to Western demands that they demonstrate the uniquely Chinese elements of their making practices. In doing so, they had been granted a certain sense of legitimacy in the international scene of making and open source tinkering. Yearnings for parity with the West and desires to achieve justice for Chinese people which I found many of China's maker advocates to variously express ironically partially reproduced a Western-centric notion of East-West difference.

The Right to Innovate

What many found so intriguing about Shenzhen's illicit experiments in electronic production was their reach into economies of scale. Shanzhai producers had quickly gained access to global markets and had done so while operating outside the international IP regulatory system. Shanzhai production had become a major mobile phone industry; within a few years, it was competing with (at times outcompeting) well-known Western tech companies from Apple to Nokia, especially in markets in Africa, India, and Latin America. It had accomplished this by producing feature phones and smartphones for people who could not afford or did not have access to the Western brands. This included phones that came in a variety of sometimes humorous shapes and sizes; phones in the shapes of Hello Kitty figurines, car keys, Chinese alcohol bottles, and cigarette boxes; pink and golden phones the size of a 5mm thin pocket calculator to fit the tiniest purse; phones the size of a brick with golden casings; phones that had a solar panel on their backs for mobile charging. It also included smartphones that were designed for specific niche markets: for migrant workers, for people of color, for construction workers, phones with cameras that capture "dark-skinned subjects" well even in low-light conditions (as such branded for the African market), phones with a dual sim card slot, phones with loudspeakers so powerful you could listen to music at a construction site. Many of these phones have become well-known brands, from the national brands Xiaomi and Oppo to Tecno Mobile in Africa.

How was this possible? Between 2012 and 2018, I was interviewing workers, designers, engineers, managers, and entrepreneurs in the shanzhai phone industry, researching its workings, transformations, and its origin stories (a thread that I will continue in chapter 6). A key aspect of shanzhai's success story has to do with the region's history of foreign direct investment (FDI), specifically outsourcing and investment from Taiwan, Hong Kong, and overseas Chinese broadly. These transnational networks, built over several decades, were key to forming tight-knit, high-trust networks of people (most of whom had moved to Shenzhen from abroad or from other regions of China) who shared resources and information within their networks to benefit each other's businesses. Early on, the industry produced electronic components and parts for industrial machinery, equipment, and instruments produced elsewhere, as well as simple devices such as pagers and mp3 players. In the years of 2004–2006, the industry shifted. In 2004, the Taiwanese chip manufacturer Media-Tek had begun producing a chip that was cheap and good enough for small portable devices. The flurry of shanzhai phones—the object that made Shenzhen's speed and economies of scale graspable—would not have been possible without this chip. The chip sets by Intel and Qualcomm in comparison were expensive. They also required engineering overhead and as such the resources of a big corporation, such as Apple and Foxconn—resources that many of the smaller industry players in Shenzhen did not have. The affordable Taiwanese chip set shifted what could be produced and enabled small, informal, illicit networks of production to proliferate.

So-called independent design houses began setting up shop. Their main business was to connect the component producers (MediaTek and the producers of other mobile phone components) with the factories that assembled the different parts into phones, tablets, smart watches, medical devices, and so on. These design houses were often run by Taiwanese businessmen, who had been operating between Shenzhen and Taiwan for several years or even decades. They produced what the industry referred to as "gongban," which literally translates as "public board"—a production-ready reference board designed for end-consumer electronics and industry appliances. The gongban public boards are designed so that the same board can be a reference (a suggestion) for many different devices; one board can make many differently shaped phones. Each gongban comes with a list of all its components and how they work together—an open Bill of Materials (BOM). This is a crucial document for shanzhai production—a document that companies like Apple keep closed. A Shenzhen-based industrial design firm basically can simply take one of the boards (gongban), keep it as is or make slight modifications, wrap it in a uniquely molded casing, and have a uniquely shaped phone. It was this process that enabled the small-batch production of a dizzying array of shanzhai phones, each batch with a slight alteration of the shape and feel of the device that directly market-tested customer interest and needs. And it was this process that attracted interest in shanzhai from both China's and foreign maker advocates. Many compared the practice of gongban to open source, but open source scaled up to mass production and global market processes. Shanzhai, in other words, demonstrated that open source could operate on the level of global mass production and trade and as such intervene in structures of such scale as global tech monopoly. It demonstrated that intervention in economies of scale was possible, something that was considered typically out of reach for an individual entrepreneur, reserved for the big industry players such as

FIGURE 3.3. Gongban (public board) for mobile phones. Photograph by the author.

Apple and Foxconn. As I outlined in the introduction of this book, one of the key promises of the American maker movement had been to prototype at vast scales. Shanzhai seemed to be doing exactly that.

I was first introduced to some of the key industry players involved in gong-ban production through David Li, who—as a Taiwanese—had unique access to it. For Li, gongban demonstrated precisely what had motivated his early work in the Shanghai-based hackerspace: to demonstrate that China offered an alternative, a more ethical approach to open source hardware and tech innovation. Shanzhai, rooted in a mundane counterculture of open sharing, was according to Li a more authentic version of democratized entrepreneurship than what he saw unfold in the networks of Silicon Valley. In the United States, he argued, open source making was an elitist practice that, despite its rhetoric of openness and democratization, largely benefited the same group of already-affluent and already-empowered players—mostly white men who were well versed in the registrars of finance capital and pitching. For Li, gong-ban exemplified how access to practices of economization and technological intervention could be truly democratized. Gongban stood in contrast to open source hardware platforms like the Arduino, which catered primarily to a fairly affluent middle class who could afford to hack in their free time. In contrast, gongban had empowered a group of people typically not considered creative at all: Chinese men, many of whom had turned themselves from migrants into successful managers and entrepreneurs through shanzhai. Li celebrated these men who did not practice "venture labor,"[71] but a more "traditional" form of entrepreneurship. He argued that this open hardware mass production of shanzhai was the "true" incarnation of the early open source ideals—the right to "hack" and thus to innovate (and to make a living by doing so) was made democratic in the figure of the Chinese shanzhai entrepreneur.

Alongside Li and Pan, another prominent advocate of shanzhai was the American hacker Bunnie Huang. Huang had made a name for himself in 2003, when he wrote a blog post entitled "Hacking the Xbox: An Introduction to Reverse Engineering," which became celebrated as one of the first publicly accessible manuals describing how to open up the "black box" of an end consumer electronic product. Media reported that because of the article, Huang had faced legal pressures from Microsoft and that MIT, where he had been studying for a doctorate in electrical engineering, officially distanced itself from him. When Huang published the manual in book form with No Starch Press, he honored Aaron Swartz who just a few months earlier had committed suicide while facing a potential USD 1 million in fines and up to thirty-five years in prison over federal charges of computer hacking. Huang did this in part to highlight the urgency of his message: "Without the right to tinker and explore, we risk becoming enslaved by technology; and the more we exercise the right to hack, the harder it will be to take that right away."[72]

In 2013, Huang began to draw connections in his writings between his conviction of the "right to hack" and shanzhai. His articulations of shanzhai overlapped in some ways with Li's and Pan's, but also differed in several aspects. For Li, shanzhai was a way to redeem Chinese production as a more ethical version of open source creativity and innovation than their Western counterparts; for Pan, shanzhai was a way to articulate making in China as rooted in a longer Chinese history of innovation that would be recognized as unique and different by the West; and for Huang, shanzhai was a cautionary tale for the West that showed how undermining people's rights to hack technology would eventually undermine Western leadership in technology innovation. He argued that reverse engineering shanzhai itself—breaking it open and exposing its inner mechanics—had the potential to recuperate the promise of open source by democratizing the right to hack and to technological ownership.

Huang's commitments to "reverse engineering" the inner workings of shanzhai in Shenzhen contributed to the public image of him as an expert, both in open source and in things China. On his blog, in speaking and consulting engagements, Huang provided glimpses into what he and others had begun referring to as Shenzhen's unique "innovation ecosystem." Indeed, Joi Ito's 2014 tour of the region with students, with which this chapter began, had worked with Huang to arrange visits to factories and to the city's infamous electronic markets of Huaqiangbei. Huang had become a face of expertise in shanzhai culture because, unlike most other foreign maker advocates, he had made (and documented on his blog) early forays into the region. Huang first traveled to Shenzhen in 2007 for a hardware start-up project he co-founded called Chumby, an early Internet of Things device that many described to me as being ahead of its time. Although Chumby was eventually considered a failure, it did not diminish Huang's credibility.

In a prominent 2013 blog post entitled "The $12 'Gongkai' Phone,"[73] Huang wrote about a phone he bought for USD 12 in a mall in Shenzhen's Huaqiangbei electronic markets. How was it possible, Huang pondered in the post, to produce a functioning phone at such a low price? Huang opened up the phone and began to trace its origins; what he discovered was the workings of gongban. Inside, he found a MediaTek chip, and, continuing to dig, he eventually found a flurry of websites "where you can download schematics, board layouts, and software utilities for something rather similar to this phone." All of that, he explained with amazement, was downloadable "for free." "I could in theory at this point," he deliberated further, "attempt to build a version of this phone for myself, with minimal cash investment." Huang claimed that all of this "feels like open source, but it's not: it's a different kind of open ecosystem." In fact, it was so unique, he argued, that it deserved a new name; "I call it 'gongkai.' Gongkai is the transliteration of 'open' as applied to 'open source.' I feel it deserves a term of its own, as the phenomenon has grown beyond the so-called shanzhai (山寨) and is becoming a self-sustaining innovation ecosystem of its own."[74] What Huang refers to as gongkai was of course the system of gongban I described earlier.

Huang here portrays shanzhai as having upgraded, moved beyond itself into a "unique innovation ecosystem," one that had "developed with little western influence." While shanzhai production cannot be disentangled from "Western influence" of course (foreign direct investment and outsourcing were key in shaping the region's production cultures), Huang ascribes to it an aura of "untouched" authenticity, a culture rich in innovation that is largely untainted by the reach of the West. "Just like the Galapagos Islands is a unique biological ecosystem evolved in the absence of continental species," he explains, "gongkai is a unique innovation ecosystem that evolved, thanks to political, language, and cultural isolation." Crucially, for Huang, shanzhai is "different from Western IP concepts" and therefore constituted "an alternate system that can nourish innovation and entrepreneurship."

A year and a half later, Huang published another post on the topic, this time entitled "From Gongkai to Open Source."[75] This is where he outlines how shanzhai might be productively "repatriated into proper open source" by reverse engineering it. This post is much longer than the previous one, going into painstaking detail how he and a friend went about "reverse engineering the Mediatek MT6260." He made clear that this project was worth the potential legal risks because a lot was at stake; the shanzhai (gongkai) ecosystem had enabled "Chinese entrepreneurs . . . [to] churn out new phones at an almost alarming pace." This was possible precisely because shanzhai operated in a legal grey zone, in which "very small teams of engineers can obtain complete design packages for working phones—case, board, and firmware—allowing them to fork the design and focus only on the pieces they really care about." As Huang put it, "As a hardware engineer, I want that." In fact, he continued, it should be our right to do so—a right that engineers were being denied by Western principles of IP and the structures of the IPR. It was not only possible to reverse engineer the MediaTek chip set, he proposed, but a moral imperative: the engineer should make the "right to hack" available to all. What was at stake in the fight for the right to hack, he argues, was comparable to the women's liberation movement and the American civil rights movement: "shying away from reverse engineering simply because it's controversial is a slippery slope: you must exercise your rights to have them. If women didn't vote and black people sat in the back of the bus because they were afraid of controversy, the US would still be segregated and without universal suffrage."[76]

Huang here aligns shanzhai with an ideal that had been propagated by advocates of the free culture movement since the late 1990s: that intervening in and devising alternatives to the antiquated structures of copyright law was an ethical and social imperative. According to the lawyer and Creative Commons advocate Lawrence Lessig, a well-known proponent of this view, copyright law is harmful because it operates on an outdated view of the world that fails to recognize the "natural laws" requiring that code itself be "free"—in other words, technology is inherently modifiable and the code should be liberated from IP laws to allow these modifications. Lessig argues that this form of "piracy" and "copying," which serves the purposes of creative production and of new value, must be understood as a form of "vernacular" creative expression in the digital age. It is therefore the right of every individual to express herself freely through this creative copying and iterating. Huang expands this liberal ideal into the context of hardware and end-consumer electronics, arguing that creative expression and innovation in hardware was harmfully constrained by the IPR that restricted reuse and rendered the copy as something inherently bad. But at the same time, Huang's extension of these ideals differed from Lessig's arguments.

Crucially, Lessig called the type of remixing taking place in Asia "bad piracy"/"bad copy," which he differentiated from acts of "good piracy"/"good copy," i.e., "creative" acts of remixing and ripping, which he associated with transformative use. In his book *Free Culture*, for instance, Lessig states plainly, "Bad piracy is Asian piracy."[77] Lessig's arguments are demonstrative of the powerful ways colonial othering endures and shapes discourses of innovation. These discursive constructs of the Asian pirate as "other," as dangerous, as mindlessly copying, as stealing rather than producing value, emerged, STS scholar Kavita Philip shows, at the turn of the twentieth century out of questions of technological authorship. The celebratory writings on technological appropriation and remix by members of the free culture movement were haunted by a pirate figure that was both distinct from and related to earlier discourses of piracy, modernity, illegality, and nationhood.[78] Like the figure of the terrorist hacker, Philip shows, the shadow of the Asian pirate lurks behind discourses about network security, threatening free markets and civilized nations and making visible the contradiction at the heart of technological progress: that the same technologies that embody post-Enlightenment modernity and progress also enable those who "hate our values and freedoms"[79] to destroy Western civilization. Anti-piracy discourse is mobilized by fear that modernity itself, with its associated sentiments of technological progress, ownership, individual freedom, and forward movement, will be destroyed. This fear of an attack on modernist beliefs and values is concentrated in Lessig's figure of the "Asian pirate."[80]

Huang's articulations of shanzhai in part redeem this very figure of the "Asian pirate." In his writings, he reinterprets Asian piracy (at least in its incarnation in shanzhai) through the figure of the tech entrepreneur. He attaches a feeling of freedom and possibility to the Asian pirate—the liberal ideal of self-realization. Shanzhai, in other words, is held up as a figure of ultimate freedom: the entrepreneur who can hack freely. This ability to hack and innovate freely was possible, Huang emphasized repeatedly in his writings, because of China's absence of laws. China here is posited as a new frontier, a place where the laws of the West don't apply, where hackers and entrepreneurs can wield technology freely (perhaps one last time).

Ultimately, Huang argues that shanzhai should be a wake-up call for the West to update its own outmoded legal structures. This was an urgent matter, Huang asserts, because the laws that had governed technology production itself were changing. In several of Huang's talks that I witnessed over the years, he reiterated the urgency of this message; the gears driving Western technological progress were grinding to a halt, and this change would have dramatic implications for the entire industry. "Technology innovation [as we know it] has hit a wall. It is the end of silicon," he highlighted. "One day in the near future, your computer won't get any faster. Your phone won't get any smaller. Your flash drive won't store more data the next year, so a lot of these things we have all taken for granted and that have driven the Silicon Valley ecosystem will come to an end." What Huang refers to here are the workings of Moore's Law. Moore's Law, named after Intel co-founder Gordon Moore, refers to the shrinking of transistors. Moore predicted that every year, the number of transistors that could fit on a chip would double and their costs would halve. However, in 2016, Intel officially announced that transistors were now only halving in size and price every 5 years. In other words, the engine that many believed to be fundamental to technological innovation—the shrinking and cheapening of transistors—had begun to stutter. In 2016, the *MIT Technology Review* called it "the death of Moore's Law"[81]—when silicon no longer shrinks at the same rate (or, eventually, at all), profit can no longer be derived from producing faster, smaller smartphones or laptop computers.

As Huang emphasized in one of his talks in 2013, the political, economic, and social consequences of these material limitations to continuous technological progress would be dramatic: "in ten years we will hit a point, when everything will have to change. Everything is going to change about everything we know."[82] But this was no reason to despair, Huang stressed; this change meant that "the best days of open hardware are yet to come." However, the industry would have to shift its conventional ideas about what counts as innovation:

> This is bad news for big companies, but it's really good news for us, the small inventors. If Moore's Law slows down from once every 18 months to once every 3 years, that gives a huge window of opportunity for someone who is doing a start-up and taking several years and developing new innovation at home. In other words, the future lifecycle of hardware will better fit small organizations, small start-ups. There is going to be more value in craftsmanship as opposed to simply a race to the bottom in specs and so forth.[83]

Huang explained that until this came to pass, small businesses could follow his lead and partner with shanzhai to approximate these conditions. Shanzhai, in other words, could be seen as helping proliferate the freedom to hack and innovate in the very moment existing norms and models of the tech industry itself seemed to be floundering.

Taken together, Pan's, Li's, and Huang's reframings of shanzhai were variously shaped by their respective positions in both Asian and Western societies; Pan's were shaped by his experiences as a Chinese entrepreneur who had given up a well-paid job as electrical engineer at the American chip manufacturer Intel to set up a maker and open source hardware business in Shenzhen that became internationally renowned; Li's by his upbringing in Taiwan, study abroad in the United States, and subsequent experience as international entrepreneur and shanzhai advocate who had declared China his home and the center of his project to redeem Chinese society; and Huang's by his positionality as a Chinese American who had "returned" to Asia (via life in Singapore) and had built a reputation as hacker-entrepreneur and early adventurer into Shenzhen. For all three, shanzhai and Shenzhen had become deeply intertwined with their respective careers and expertise. They had become variously vested in recuperating the (male) figure of the Asian pirate and in redeeming Chinese people as innovative and creative. They each had modeled ways in which the Western tech industry could (and should) engage with China differently. And they each had become widely recognized as experts in things China, innovation, entrepreneurship, economies of scale, and future making. And it was their articulation work that laid some of the crucial foundations for how China would eventually be rescripted in the broader tech imagination and for how shanzhai was taken up in the worlds of design, consulting, architecture and urban planning, and education.

Designers in Touch with the World

> [Shanzhai] has this incredible ability of developing a sense of touch. We high-tech nations are clueless how to do most of this kind of stuff . . . we tend to underplay the value of this kind of thinking, i.e. embodied thinking.
> —JOHN SEELY BROWN, INTERVIEW WITH THE AUTHOR, CHINA, 2014

In 2011, the Shanghai office of the American design consultancy IDEO took up the idea of XinShanzhai (new shanzhai) as articulated by China's early open source hardware and maker advocates. IDEO began hosting its own version of the event, titled "XinShanzhai: An Open Platform for Business Innovation." Soon after, the American consulting firm published a guidebook on "designing for shanzhai" that stipulated,

> Countries, from the US to Japan, regularly accuse China of copying designs. Indeed, multinational companies in these countries spend an inordinate amount of time and money trying to prevent their products from being copied. But Shanzhai—"copycat" design—represents a vast business opportunity. Shanzhai is an open platform for grassroots innovation . . . shanzhai designs are an opportunity for international companies to introduce Chinese consumers to their brands, and then observe how local Chinese culture adapts their offerings.[84]

The booklet lists several steps for how (Western) companies might "harness learning by observing shanzhai design": "leveraging the wisdom of common folks," "using tools for expression (that enable people to mark a bit of themselves on their products)," "going for liangdian or 'shiny points' (being clear about the value proposition of your product)," and by "exploiting grassroots sentiments (harness grassroots humor to get closer to Chinese consumers in diverse regions)." In this booklet, shanzhai is presented as providing a unique "platform" for "foreign corporations" to learn how to adapt specific products to the Chinese market, or to target Chinese consumers. Ironically, IDEO itself had allegedly been struggling to enter the Chinese market,[85] and their shanzhai booklet became a playbook for IDEO itself. The company rebranded its own consulting business in Asia by translating shanzhai into a business advantage for Western corporations.

In the worlds of design and tech consulting, people turned to shanzhai as a pathway for companies to adapt to a shifting global market—as a business opportunity for struggling Western companies. Shanzhai here was seen as "illuminating for foreign managers looking for ways to improve their own business model," as the design consultancy Booz&Company claims in a widely circulated 2012 report.[86] According to this report, shanzhai was the distillation of the "Chinese mentality" of "fearless experimentation"—the mentality that had given Chinese companies an "advantage over slower-moving foreign competitors." In the report, China's ability to use Western tools of innovation faster and better to beat the West is compared to a concept originated by Clayton Christensen of the Harvard Business School, the "innovator's dilemma."[87] According to Booz&Company, shanzhai and China itself, "boosted by the fearless experimenter mind-set" of shanzhai, were improving on the familiar Silicon Valley model of disruptive business. The company urged foreign competitors to imitate China, thus framing the Chinese copy of Western business models as a prototype, a model for the new business direction needed in the West: "get out of your comfort zone. Learn from the Shanzhai players. Become a fearless experimenter." This idea of shanzhai as a model for Western companies competing in a fierce global market was adopted by a growing number of business and design consultancies, which saw shanzhai as embodying a moralized capitalist logic of creative destruction driven by survival instincts and an experimental disposition, not by privilege or prior success. Shanzhai thus appeared to be capable of revitalizing Western businesses.

As a growing number of international business magazines, scholarly publications, and arts and architecture projects and institutions took an interest in shanzhai, examining how it challenged Western-centric notions of modernity, creativity, and progress, it came to stand for the ultimate incarnation of the postmodern sensibility, of the suspicion of modernist ideals of rationality, progress, and authorship. This was brought home during the Shenzhen Bi-City Biennale of Urbanism\Architecture (UABB) in December 2015, with several panel discussions and exhibitions evolving around the themes of Shenzhen, shanzhai, and design. One such exhibition was put on by Louisa Megnoni and Brendan Cormier of the Victoria & Albert Museum, entitled "Unidentified Acts of Design." The exhibition featured an assortment of artifacts taken to demonstrate Shenzhen's unique approach to design. These unique and promising artifacts were "unidentified," i.e., unbranded and with no author(s)/creator(s) named, and therefore—as the exhibition made clear—unseen by the West.

The curators aimed to correct the West's unseeing of these artifacts, and the UABB, with its large numbers of foreign (mostly European) attendees,

FIGURE 3.4. "Unidentified Acts of Design" exhibition at the 2015 UABB (Shenzhen Bi-City Biennale of Urbanism\Architecture) by Louisa Megnoni and Brendan Cormier, Victoria & Albert Museum. Photograph by the author.

posed an ideal context in which to do so. The exhibition showcased a range of objects, including "anonymous" shanzhai phones, but oddly, it also included quite *identified*, even branded artifacts: such as a DJI drone, swag from the Chinese Internet company Tencent, a robotics kit from the company Makeblock, and more. In a 2014 blog post leading up to the exhibition entitled "researching *Unidentified Acts of Design*," Cormier explained the exhibition's aim:

> Unidentified Acts of Design is an exhibition and research project that seeks out instances where design intelligence has occurred in Shenzhen and the Pearl River Delta outside of the conventional notion of the design studio. The project aims to show how in a region of unprecedented growth, which has long served as the factory of the world, design acts can take on unconventional forms and occur in unpredictable places. By seeking out new definitions of what constitutes design, new actors and new objects are able to enter into the canon of the region's design history, while an expanded sense of design's relationship with the region can take shape.[88]

Shanzhai, the curators suggest here, helps the designer to "seek out new definitions of what constitutes design" "outside the conventional notion of the design studio." This is a rearticulation not only of shanzhai (which is assigned

FIGURE 3.5. Shanzhai phones featured in the "Unidentified Acts of Design" exhibition at the 2015 UABB. Photograph by the author.

new meaning in this context) but also the role of the designer. By moving "outside the conventional notion of the design studio," the designer can open up the "canon" of design history, an action that in itself is understood as a more ethical—because more inclusive—approach to design. Note that this "naming" of a renewed, ethical design approach ironically was made possible because, in relation to the West, shanzhai was rendered "unnoticed" and "unnamed." I can't help but see in the museum's project a reminder of the colonial project, which appropriated otherness—here, the collective spirit of invention and production in Shenzhen that does not depend on IP and branding—in order to advance a Western ideal and a Western institution—the idea of individual achievement and individual creativity concentrated in the figure of the wordly designer.

Following the exhibition's official opening at the UABB, I watched a bewildered-looking Chinese man walk up and down its display of shanzhai phones. Curious, I inquired what he thought of the exhibition. He laughed at first, saying it was "interesting." We talked about the various objects represented until he got more serious: "you know—well, I want you to know," he said quietly, "my company made one of these phones. It is not unnamed at all." I took the man's name card and promised to follow up with him. He had been working in Shenzhen's phone business for more than ten years. When I returned to the exhibition later that day, I noticed that the man had left his business card next to the phone he claimed his company had made. The small, subversive act made me smile, for it made visible, at least for a moment, how the designer's orientalist embrace of shanzhai as "unidentified" had masked so much of shanzhai's actual workings.

In 2018, a version of the V&A's exhibition at the UABB moved into Shenzhen's new Design Museum, "Design Society (设计互联 sheji hulian—literally: interconnected design)." The museum itself was envisioned to elevate Shenzhen's modality of experimentation as key to "reformulating" design more broadly. "It's a museum, but of a different kind," Ole Bouman, the museum's director, told me in an interview I conducted with him in 2016. When I asked why Bouman thought that he, a foreigner, was hired as the director of the city's first design museum, he explained,

> I think part of the reason I am here [as the director] is to help articulate [the museum] . . . The articulation is part of my role . . . for instance the name Design Society is a perfect example of this articulation. So, from the original idea to set up a design museum, now we set up a design society. It's more like an agenda rather than a place. So, the original name was a place, bowuguan [Chinese for museum] and then there was society as an agenda for positioning design in new ways . . . we thought this would be much more out-reaching and societal and that's why we came to this, the other name.

Bouman here positions the museum's task as articulation; it is "an agenda" rather than a "place," which would produce a "design society"; in other words, the museum's name and its object point to the notion that a design museum in Shenzhen can create a new society or transform existing institutions. Being the director of such a museum, Bouman explained, required that he "immerse" himself, "observe," and learn from Shenzhen how to be "humble" and to confront his own biases. It was through this process of experimenting with an alternate subjectivity, he made clear, that one can "see the potential of the city," learn from its "practical mindset," and "feel the [city's] drive, to use that energy, to survive, not only to discuss academically, but to live, to survive."

This idea that learning from Shenzhen (and from the "fearless experimenter mindset" of shanzhai) would enable the designer to prototype alternative approaches to design and *in doing so recuperate the designer as moral and humble* was articulated in various corners of the world of design from consulting to architecture, art, and urban planning. While shanzhai became understood as a copy that allowed the restoration of the designer's morality in the very moment as debates over design's complicity in various forms of exploitation and exclusion were peaking and calls for design justice[89] and for rethinking design were proliferating, Shenzhen was construed as the laboratory within which to craft this new image. It presented, in other words, not only an opportunity for the cosmopolitan designer, but also for China, to achieve the status of a happy, modern nation (a narrative that appealed to the CCP, see chapter 6). At the 2016 Shenzhen Maker Faire, Bouman articulated this idea publicly on stage. His talk began with what has become a somewhat iconic phrase associated with Shenzhen, "来了，就是深圳人," which loosely translates as "If you come, you are a Shenzhener."[90] The phrase flashed across the screen behind Bouman, depicting a young Chinese man shouldering a backpack, with the Shenzhen skyline behind him in the near distance. It was the same image I had seen before, plastered on walls at the Shenzhen airport and subway stations as part of a promotional city campaign by the municipal government. "This image of Shenzhen—the appeal, the magnetism of Shenzhen is what I want to talk about," Bouman declared. "Probably all people in this room know this poster. I don't even know how old this poster exactly is. But it says something about the city as an attractor, as an engine of growth, as an emancipator. The power of this image says something about China's future, which is about creativity and about taking risks . . . it says that if you are smart, if you have the courage, it will happen to you, and you will be the main driver of this."[91] Shenzhen was fundamentally about making things, Bouman argued, "we can't imagine another image of Shenzhen than as the maker city." But this was not the case for Chinese culture broadly, he cautioned. Thus, following this image of Shenzhen, he continued, was paramount for China, "this image of Shenzhen as a city to become happy, a city where life is good . . . this magnetism of Shenzhen . . . has enormous potential . . . of course, you are not just happy, you are happy only if you do something with your freedom, with your Shenzhen identity, when you start to make things." And this was exactly what Design Society was committed to enabling: "the Shenzhen Design Society is a place to make this transition happen," Bouman ended his talk, "to enable [Chinese] people to move from making things to become real creators, and to become uniquely and globally relevant that way."[92]

Genuine Arduino "Made in China"

The broadening appeal of shanzhai unfolded alongside the eruption of a series of controversies in the world of open source hardware and making. Shanzhai became the center of one such heated debate over the broken promises of open technology and the maker movement—what became known as the "Arduino wars."

FIGURE 3.6. Massimo Banzi and Eric Pan introduce the "Made in China" Genuino board, 2015. *Make:* magazine.

In June 2015, at Shenzhen's second featured Maker Faire, Eric Pan of Seeed Studio and Massimo Banzi, the CEO of Arduino, announced that Arduino and Seeed Studio would partner on the production of a new open source hardware platform, the Genuino—the first "made in China" board officially sanctioned by Arduino. The Genuino was introduced with much fanfare, as Arduino had done with its boards in the past. In the photos that appeared during and after the Maker Faire, Banzi and Pan were shown side by side, as they made their deal official.

Despite the cheerful presentation at the Faire, the release of this "made in China" board was potentially risky for Arduino. Arduino's identity had long been tied to its "made in Italy" branding, which the company said distinguished it from so-called Chinaduinos, cheap clones, derivatives, compatibles, and counterfeit Arduinos "made in China." On its website, Arduino had detailed, for instance, "how to spot a counterfeit Arduino," recognizably different, or "other," from the carefully crafted design of an authentic Arduino board: original and copy could be distinguished by paying attention to the color of the PCB silkscreen, the font of the logo, the color of one of the components ("a major fake alert"), the "pasted" and "ugly" connectors and connections between counterfeit components, as compared to the "woven" and "beautiful"

connectors of the original.[93] Another clue was, of course, the price, for counterfeit Arduinos were around half the price of the original. The final key difference according to Arduino was "the map of Italy, saluting the place of its
birth," printed with sharp contours on the original and which was, by contrast,
often hazy or altogether off in the counterfeit. Arduino and "made in Italy"
here stand for craft, design, authenticity, beauty, and value, unlike the "ugly,"
"fake," and "low priced" Chinaduinos.

Banzi had stressed this view not only on the official Arduino blog and website, but also in some of his public lectures. At the 2013 Hardware Innovation
Workshop in San Mateo, for instance, he emphasized that it was his company's
Western design heritage that differentiated an authentic Arduino board from
the copy. Banzi's talk invoked several design traditions, from the Bauhaus (with
its ideals of cultivating individuals simultaneously as "more completely indi-
viduated *and* more integrated into society")[94] to principles of human-centered
design, design thinking, and open source. In Banzi's words, Arduino had created "not a product," but "an experience": indeed, it was its commitment to
user experience design that had enabled Arduino to develop its brand, which,
he argued, was profitable precisely because it stood for open-ness, beauty, and
democratization:

> Arduino is basically a mash-up of open technologies, so there is a bunch
> of different technologies that we put together and we wrapped it up in an
> experience. I think that's important for us, the user experience, the out of
> the box experience, the speed—how long does it take for you to get from
> zero to something that is working. That time is very important. The longer
> it is, the more people you lose in the process. And then we looked at how
> you can deal with open source, how to keep everything open and protect
> the brand . . . We took inspiration form the fashion industry. The fashion
> industry doesn't have all the protection that the movie industry has, but it
> makes much more money, and still they can't trademark this shirt, because
> it's all about the brand and the relationship to the brand.[95]

Banzi here aligns branding (which rests on notions of corporate profit, IP
enforcement, and marketing) with what is often understood as its opposite: the
ideals of open source. The role he assigns to "user experience" is more reminiscent of the corporate branding strategies deployed by Apple than of the values
of the open source and free culture movement. Just as Apple claims to sell not
only phones but lifestyles and consumer identities, Banzi claims that Arduino
sells not simply open source hardware boards, but an authentic open source
experience. Arduino, in other words, stood for doing open source hardware
"right." As Banzi summarized it in the key phrase of the talk, Arduino was
not about hardware, but about people: "pay less attention to megahertz but
more attention to people." It was this attention to *people* rather than *hardware*

that allowed Arduino to claim a kind of moral superiority in the open source hardware community and to frame as morally suspect counterfeits, copying, and other acts of reproduction.

Given this careful branding of Arduino as "made in Italy" quality, rooted in design principles of user experience design and traditions like the Bauhaus, why would the company decide to produce one of its boards in China, and why did it partner with Seeed Studio to do so? As I show in what follows, the decision had much to do with Arduino's attempts to restore its claims of authenticity in light of a controversy in the open source hardware community in which the company had become deeply embroiled.

About five months before the "made in China" Genuino board was announced, news had spread that the founders of Arduino were enmeshed in an internal dispute: on one side was Arduino LLC (which was owned by four of the Arduino founders: Massimo Banzi, David Mellis, Tom Igoe, and David Cuartielles), and on the other side, Smart Projects (owned by the fifth co-founder, Gianluca Martino). Smart Projects was Arduino's Italian manufacturing arm, and it was thus crucial for Arduino's "made in Italy" brand identity. In 2014, Martino sold Smart Projects to Frederico Musto, who renamed Smart Projects as Arduino SRL and established a new website, arduino.org.[96] The original company, Arduino LLC, subsequently sued the new Arduino SRL for trademark infringement. The lawsuit made visible something that until then had received little attention: the fact that Arduino's community-facing entity (Arduino LLC, with Banzi as the main figurehead) was organizationally and legally separate from the company that actually manufactured its boards (Smart Projects, run by Martino).

While the internal lawsuit focused on trademark infringement, the disagreement between the co-founders was allegedly rooted in fights over Arduino's global production strategy and business model. In 2013, Arduino LLC and the American chip manufacturer Intel had announced a new partnership, agreeing to "work closely together on future products that bring the performance, scalability, and possibilities of Intel technology to this growing community of makers." At the 2013 Maker Faire in Rome, Banzi released the Intel-branded open source board Galileo as an "Arduino certified board featuring Intel architecture." All of this indicated that Arduino LLC was serious about exploring manufacturing partners other than the Italian-based Smart Projects. In other words, Arduino LLC had for the first time begun to officially disassociate its brand identity from "made in Italy," moving toward potential independence from its original manufacturing partner, Smart Projects. Maker-related media in Europe and the United States later speculated that Smart Projects must have seen that their revenue model, which was tied to Arduino LLC, was threatened by the shift in branding and the move toward globalized manufacturing. Indeed, it was reported that Smart Projects had renamed the company and bought the domain Arduino.org in a bid to claim ownership in the company and reserve a say in the company's future manufacturing strategies.[97]

By partnering with Seeed Studio, a Chinese company that had become widely celebrated as a "true" open source innovator in China, Arduino was able to retain its moral position, simultaneously distinguishing itself from the "made in China" counterfeits and from its former "made in Italy" partner, now accused of trademark infringement. Genuino, a clever play on the word "genuine," framed the "made in China" board produced with Seeed Studio as being just as "true" to open source as the "made in Italy" boards. When Banzi announced the partnership with Seeed Studio, he emphasized as much: "the new Genuino name certifies the authenticity of the boards, in line with the open hardware and open source philosophy that has always characterized Arduino." The partnership with Seeed Studio, in other words, allowed Arduino LLC to demonstrate that it continued to follow open source ethics and values, even as Arduino was giving up what had been one of its core claims to originality, quality, and authenticity: the "made in Italy" brand (its manufacturing location in Italy). Banzi explained this in an interview with the Italian press:

> Arduino is an open hardware project: anyone can make it, projects and schemes are online. What differentiates the original version, so to speak, is its identity. Being at the center of the maker's movement for ten years. It is a brand because it has its own philosophy and a history. Pure and simple hardware is not the center. There are several factories that make Arduino boards around the world. However, that of Ivrea, which belongs to Gianluca [Martino], has historically had an important share in production. My idea of expanding into the world clashes with his fear of having to reduce production. But if you really want to land in China, where they already copy Arduino even with our logo, you cannot do it by continuing to keep most of the production in Italy.[98]

Banzi argues here that the value of Arduino was its decade-long place at "the center of the maker movement" along with its user experience design, while the "pure and simple hardware" had always been the site of mere execution. Moving production to other regions, in other words, would not taint Arduino's core identity or ideological commitments. Banzi also invokes the Chinese copy in order to further legitimize the company's shift to non-Italian production sites; it was the "copy Arduinos," he claims, that had forced Arduino LLC to disassociate itself from its manufacturing partner in Italy, for Chinese shanzhai production offered cheaper (competitive) products. Shanzhai appears as a villain in Banzi's story, as does the Italian manufacturer portrayed as at fault for its inability to keep up. In this narrative, Arduino LLC itself is not accountable, for it was merely responding to the natural progression of market expansion via flexible accumulation and displaced production.

The figure of the Chinese copy was part of Arduino's narrative already earlier, when the company explained its decision to trademark Arduino in the first place (the very decision that would animate the "Arduino vs. Arduino" court case). In a 2013 blog post entitled "Send in the Clones," Banzi explains that Arduino's decision to register a trademark was motivated by the Chinese copies, for the company "needed a way for people to be guaranteed that they were buying a quality product." It had decided therefore "that the best way was to register the trademark of the Arduino lettering and to create a logo that would make it easier to identify products sanctioned above." Banzi argued that Chinaduino "counterfeits"—"boards that clone the official board including the Arduino branding (logo and board graphics)"—were "really detrimental to the whole open-source hardware movement." The decision to trademark, as Banzi explained, was motivated not by profit but by the company's ethical commitments to protect the user from being wronged or harmed by these counterfeits: "since the Arduino graphics are trademarked and we don't release any of the files, whoever uses our graphics and logo makes a deliberate act of trademark infringement. These products not only trick people into thinking they are buying an official Arduino (therefore supporting the Arduino project) but they also provide no support." These classifications of clones and counterfeits versus original claimed a position of relative morality by framing the counterfeits as "tricking people" and "stealing identity" and therefore failing to "contribute to the Arduino project" and the "open source hardware" project.

Once the lawsuit came to light, setting off what some referred to as the Arduino Wars, heated debates unfolded online. While some sided with Arduino, there was a broadening sense that the Arduino Wars were just another example of how the ideals of the maker movement were being co-opted by the very forces it had aspired to undo: the furthering of corporate monopoly, technological elites, and Western-centric notions of design. Some pointed out that the high prices of the official Arduinos made them inaccessible, shutting potential users out of hardware, and that the clones were far more accessible. Others shot back that this attitude was short-sighted because it ignored how the "OS model" would benefit the whole community. But many people agreed that the expansion from "Arduino to Genuino" was absurd, and the announcement was greeted online with ridicule ("the most ridiculous thing I have ever heard" and "Massimo Banzi is Arduino!"). The announcement created months-long online debates about the ethics of open source and the distinction between copy and open source. When the Arduino co-founders appeared to settle their dispute outside of court in 2016, Hackaday, a widely read news blog in the open source hardware community, declared, "the Arduino wars are over. Arduino is dead, long live Arduino." Indeed, little more could be said, since questions of ownership were resolved behind closed doors. Yet the various responses to the Hackaday post made it clear that the lawsuit and resulting controversy had tainted Arduino's image, at least for some. Others invoked racial and cultural stereotypes, arguing that the dispute and the continued secrecy had little to do with open source but more with "Italians being involved." And yet others argued that they had long since given up on purchasing Arduino, as it "is no longer the best in any category" and that "most people now buy Chinese clones." According to some Hackaday commenters, "the real creator of Arduino" was not any of the "Arduino guys" involved in the dispute, but a sixth person: Hernando Barragan, who—as a commentator explained—was given zero credit for his innovation Wiring, which was where "most of Arduino originated from." The same commenter argued that "from a moral and ethical perspective, there would be much to discuss, as right or wrong depends on one's (or a community's) beliefs and rule of conduct."

There had been many previous online debates about the ethics of various open source hardware projects. For example, there were open questions of how open source hardware related to (or should relate to) commerce and to the military. Between 2012 and 2013, the previously open source 3D printing company MakerBot decided to go closed source in order to open itself up to investment; in the same period, Maker Media announced that they would accept funding from the US military agency DARPA in 2012. Shanzhai and China were central protagonists in all of these debates. Shanzhai was seen as both a villain (the reason for the fall of open source hardware) and the hero (as a hopeful alternative to Euro-American-centric notions of creativity and ownership). Shanzhai thus directly and indirectly shaped the conversations and attitudes toward Arduino, for the morality of open source had come to hinge on how the Chinese copy was to be understood. It was at the height of the Arduino Wars that more open source advocates, designers, and engineers began to look favorably upon shanzhai and Shenzhen; the controversy over what constituted moral open source and tech innovation writ large began to shape a new imaginary about the copy and Shenzhen.

Free culture and public domain advocates used "'creativity' [as] an axis through which to differentiate the good copy from the bad copy,"[99] leaning on a Lessig-type moral discourse of creativity that sought to distinguish piracy from transformational use and improper copies from proper copies. Although free culture and public domain advocates see themselves as antagonistic to the IP regime, they share with copyright fundamentalists an ontological common ground: modernist ideas of technological authorship as a function of individual creativity, transformative production, technological change and progress, and, ultimately, capitalist productivity. The binary split between "good" and "bad" copies runs along the fault lines of economic and technological progress, "reproduce[ing] the hierarchies that animate IP itself";[100] according to legal scholar Lawrence Liang, piracy was cast as the villain in this morality play because (unlike public domain and open source) it did not market itself as being part of a narrative of resistance or appropriation. In other words, piracy, which "does not stake claims in the world of creativity," is therefore seen as "operating within the logic of profit and the terms of commerce" and thus "cannot claim the same moral ground available to other nonlegal media practices."[101] Shanzhai—as a Chinese pirate figure—emerged in the transnational tech imagination in the precise moment that the modernist ideals of individualism, creativity, and progress that had been used to distinguish Asian piracy from the public domain were increasingly controversial.

"Time to Copy China"

In 2016, *Wired* magazine ran a cover story whose headline asserted "It's Time to Copy China." It featured Lei Jun, the CEO of the mobile phone company Xiaomi, which has grown out of Shenzhen's shanzhai industry. In the issue's editorial article, David Rowan claims that "Xiaomi's $45bn formula for success" was not simply that it copied Apple. Instead, it had successfully transformed the smartphone business by collaborating with smaller enterprises and corporations—what would be seen in the West as their competition—as allies and partners. According to Rowan, this model was so successful that it should be copied by others:

> Xiaomi makes only smartphones and tablets, TVs and set-top boxes, and routers; everything else is produced by independent companies in which Xiaomi has invested between $100,000 and $500,000, from the smart blood-pressure monitor (made by California's iHealth Labs) to Yunmi Technology's water purifier. "We sell 600 items," Liu [Xiaomi's head of "ecosystem products"] says. "If we did this alone, we'd need 20,000 people in the company, but we have only 8,000. We've reviewed 600 startups in the past two years, and invested in 54. We help them define their product and use our sales channel, supply chain, branding and financing. They're our special forces, and we are the commander."

This article presents Xiaomi as being in tune with the ethos of open innovation and even open source. Rather than undermining competing ideas through expensive lawsuits, it had successfully implemented a business model of shared profit and partnership, with each entity focusing on what it was uniquely good at. According to Internet theorist Clay Shirky, Xiaomi was "not a repudiation of Shenzhen's shanzhai culture," but "its next form": "Xiaomi has shown how to take relatively commoditized parts and turn them into a desirable product, through a combination of physical design, continuous improvement of software and services, and something like an aura."[102]

Shirky's mention of "aura" is particularly interesting. The term was made popular by Walter Benjamin's often-cited essay on "The Work of Art in the Age

FIGURE 3.7. Cover of *Wired* 2016: "It's Time to Copy China." On the cover page is Lei Jun, CEO and founder of the Chinese mobile phone company Xiaomi.

of Mechanical Reproduction" ("Das Kunstwerk im Zeitalter seiner technischen Reproduzierbarkeit"), which argued that reproduction—in other words, *copying*, whether "good" or "bad"—devalued the aura of an original piece of art, which was grounded in its authenticity. Yet Shirky connects the "aura" of the original to mass-produced Xiaomi phones—mass production imbued with an affect of entrepreneurial agency and morality. Shanzhai here is seen as intervening not only in the monopolies of IP regulation but also in the West's "auratic traditions of 'true art'"—and its modernist ideals of authenticity, originality, and individuality.[103]

I have shown in this chapter how shanzhai, which disassociated innovation and creativity from Western notions of the individualized inventor and creator, was seen as simultaneously threatening and offering an alternative to masculinized and individualistic notions of design and innovation.[104] While this view of shanzhai was at first controversial, the debates that ensued laid bare the contradictions of the moral discourse around open source and peer production.[105] Although making is often seen as fostering new communities and new forms of collaboration and collective action that work outside of (and potentially against) formal institutional structures, it was founded on ideals of individualistic creativity. It was these ideals that were the subject of debate at Maker Faires, in WeChat groups, on Western social media, in blog posts, and eventually, in the mainstream media. As I have shown, several controversies in the open source community about authorship, ownership, and innovation catalyzed the re-articulation of shanzhai, seen as affecting open source in particular in two contradictory ways: both as having the potential to recuperate open source hardware as the ethical field of technology it claimed to be *and* as being a threat to the very foundations of the project of open source. These confrontations cast doubt on the seemingly firm definitions of "good" design and innovation. "Learning from the fearless shanzhai experimenters" was construed as enabling people to redefine design and innovation. It promised the prototyping of a new kind of professional identity in design, open source, and engineering, one that rekindled the belief in technological promise, hope, and futurity. This professional identity prototyped in Shenzhen differed significantly from previous notions of what constitutes the ideal design and engineering subjectivity—the ideal of the removed, distant, and rational thinker, a white-collar worker with clean hands, sitting at a computer in a studio or office, unencumbered by the economic and political realities that make digital infrastructures work. In contrast, the Shenzhen prototype of the professional was eager to learn from the world and immerse himself in its day-to-day workings. He engaged with the messy reality of labor issues and geopolitics. He reflected on his own bias, performing a version of feminist critique by showing how Western knowledge work ignored its dependency on material labor and labor exploitation.

However, he did not examine all his biases. I deliberately use the male pronoun here, because while this renewed hacker, designer, and engineer challenged established divisions between East and West, other and self, individualism and collectivism, creativity and copy—in other words, the entire telos of modernization in the West, he often overlooked the fact that the engineering and design fields have gendered and racial inequalities, often justified by the claim that these types of expertise are difficult to democratize.[106] The interest in shanzhai co-evolved with a broader endorsement of feminist critiques, although these were seldom acknowledged in those terms. The uptake of critical sensibilities via Shenzhen and shanzhai allowed people, many of them men with power in established design and technology networks, to re-establish technology production and design as a virtuous enterprise[107] and prototype a renewed ethical and moral designer and engineer self. These prototype professionals thus built alternative subject positions. But they reserved these new subject positions largely for themselves and those associated with the Western institutions and corporations they represented, from the MIT Media Lab to Arduino. These prototype professionals were able to re-invent themselves as "ethical hackers," because a strong discourse of digital counterculture and disruptive innovation supported them, as long as the structures of investment and the ways of Silicon Valley were still respected.[108]

As I am working on the final edits for this book amidst heated controversies over China's telecommunications giant Huawei, fears over a China-US trade war, and American media portrayals of China as a new technology superpower engaging in police-state levels of surveillance capitalism using AI and machine learning, the notion that China would *not* centrally shape the future of the global tech industry seems absurd. But just a few years ago, things were different. When maker and open source advocates, entrepreneurs and designers, architects and engineers were flocking to Shenzhen, China was hardly understood as a forceful and assertive actor in global economies of technology production. Although China has long figured in the Western imagination through narratives of threat and fear from the "red scare" after World War II to the "rising dragon" in the 1990s, it has rarely been seen as "being in charge" of how modernity, alternatives to the West, and technological progress were defined and implemented. One of the underlying goals of this chapter has been to unpack the role displacements of technological promise play in how China is perceived globally and in shifting China-US relations. As shanzhai garnered attention from a wide array of actors, their articulations of copy, open source, quality, design, fakes, counterfeits, and IP helped to shift attitudes toward China both in the Western media and in the broader international tech imaginary. Shenzhen was suddenly seen as a central player in the markets of future making: its "otherness" and backwardness were now celebrated for their difference, and its unique "features" were marketed as "opportunities" for entrepreneurial activity and the recuperation of technological progress. Orientalism, here, expresses itself in what Arif Dirlik called the rise of "various nationalisms" that assert the East's difference from the West in order to compete for investment in a global market of finance speculation.

The displacement of technological promise onto regions once labeled backward and considered the tech periphery serves both powerful political elites and the machineries of capital investment as they capitalize on the promise of difference and regional advantage. But displacements of technological promise don't simply happen; they have to be actively produced and articulated. As I have shown in this chapter, our own scholarly embrace of certain sites or practices as hopeful, as innovative, or countercultural participate in and coproduce these displacements. The scholarly excitement about various forms of hacking as carrying in and of themselves political potential and the scholarly intrigue with regions that seem to offer a grasp of scale and speed in action are not innocent. The endorsements of certain practices, sites, and regions as offering an alternative and hopes to rediscover (some form of authentic) resistance (in the small and in the large) in spite of the neoliberal mantra that "there is no alternative" to capitalism co-produces the difference that investors seek and that governments desire so they can brand certain regions as carrying renewed regional advantage. We might ponder, at the end: just how radical are endorsements of the broken worlds of modernity?

4

Incubating Human Capital

MARKET DEVICES OF FINANCE CAPITALISM

Making today is two clicks away from becoming an entrepreneur. It's about taking the steps from self-reliance to self-employment.

—BRE PETTIS, "WHY EVERYONE IS A MAKER NOW"

Human capital shifted the iconic economic subject from a worker or a consumer to an entrepreneur whose body contained an alterable set of assets and risks, a reknitting of *homo oeconomicus*.

—MICHELLE MURPHY, *THE ECONOMIZATION OF LIFE*, P. 115

As investors in their own human capital, the subjects that are presupposed and targeted by neoliberalism can thus be conceived as the managers of a portfolio of conducts pertaining to all the aspects of their lives.

—MICHEL FEHER, "SELF APPRECIATION," P. 30

Mike Wang was an early member of the Shanghai-based hackerspace XinCheJian.[1] In 2013, he made the news in the United States; Wang, who had started pursuing a degree in electrical engineering and computer science at UC Berkeley just a year earlier, had received a "Thiel fellowship." Peter Thiel, co-founder of Paypal and early investor in Facebook (and more recently in the news for the furor his endorsement of Donald Trump caused in Silicon Valley), set up the fellowship in 2010. Fellows must be younger than twenty years old. In exchange for dropping out of college, they receive USD 100,000 to enable their entrepreneurial pursuits. Thiel (who himself has a BA in philosophy and a law degree from Stanford) is well-known for criticizing higher education. He gave his views in a book he co-authored with David Sacks, *The Diversity Myth: Multiculturalism and the Politics of Intolerance at Stanford*.[2] The Thiel fellowship offered to Wang is one of many alleged alternatives to higher education that have sprung up over the last 10 years: tech incubators, accelerators, hackathons, start-up weekends, and adjacent entrepreneurship and start-up training programs. These programs feed off and fuel the argument that higher education is unable to prepare young people for the economic realities of our times—a critique that has become more pronounced in the United States since the financial crisis of 2007–2008.[3] As Thiel said in 2014, "We need to create a much more diverse array of alternatives [to college] . . . ," and "there is no reason why the future should happen only at Stanford, or in college, or in Silicon Valley."[4]

Programs like the Thiel Fellowship aim to "disrupt" education by providing a competing platform: tech entrepreneurship training in the tech incubator. This training is pitched as democratizing tech innovation, i.e., as making innovation accessible to "everyone." It is the promise that democratized entrepreneurial life will spur national economic development. In his 2012 book, *Makers: The New Industrial Revolution*, Chris Anderson argues that incubators are "start-up tech factories" and "start-up schools" that "coin entrepreneurs first and ideas later."[5] Echoing Thiel, Anderson argues that the only education that mattered was the kind that served economic development via entrepreneurial agency—education that did not promise a job but instead trained people in building their own jobs in a global market of finance speculation. Incubators train a particular kind of entrepreneur—they train people to see themselves as entrepreneurs of themselves, as investors in themselves.[6] Tech incubators invest in ideas, but more important, they invest in the promise of "nurturing" into being a particular kind of innovator.

Most incubator and adjacent tech entrepreneurship training programs are tied to venture capital investment or large corporations. Ironically, universities, colleges, and high schools—the very institutions that tech incubators are designed to disrupt—have increasingly opened their own entrepreneurship programs, incubators, and innovation hubs, where they run hackathons, accelerators, start-up weekends, and maker competitions. These programs, as the information and STS scholar Christo Sims argues, promise to "sweep away antiquated educational conventions and replac[ed] them with an innovative and improvisational culture that was more akin to a Silicon Valley startup than a traditional public school."[7] These initiatives, celebrated as models of "engaged learning" that take students out of the classroom and into the real world, are used by the university administration's development and marketing offices to promote the university as competitive in the lucrative market of higher education. In these "start-up schools," students are no longer consumers of higher education. Instead, they are producers, trained to self-invest, nurtured (like the stock of financial markets) to increase the capital to which they are attached. Incubators produce a particular kind of subjectivity; as noted above, they train people to become investors in their own human capital.[8]

Human capital, political theorist Wendy Brown suggests, "is the next step of *homo oeconomicus* as a neoliberal agent that seeks to strengthen his/her competitive positioning. Neoliberal rationality remakes the human being as human capital."[9] Similarly, the historian Michelle Murphy argues that human capital is the "economic subject shifted," and this shifts people "from a worker or a consumer to an entrepreneur whose body contained an alterable set of assets and risks, a reknitting of *homo oeconomicus*."[10] But how exactly is this human capital cultivated? What explains its broad uptake, its affect, its lure? Why would people be eager to become investors in their own human capital?

One of neoliberalism's great myths is that market capitalism is laissez-faire, that competition and the "free market" will regulate themselves with no intervention. In this chapter, I show that human capital is all but a natural outgrowth of economic development. The subjectivity of self-investment and human capital has to be produced, nurtured, incubated, and trained into being. I speak to the historical condition that gave rise to this process of subject-making in the tech industry and to the instruments, devices, methods, and techniques that enabled it. I draw from ethnographic research that I conducted in 2013 with the founders, participants, and employees of a hardware incubator program in Shenzhen, run by a Western venture capital and investment management firm heardquartered in the United States. The chapter is also informed by hundred and twenty interviews with founders, investors, program managers, and participants in similar programs in China, the United States, Taiwan, Singapore, Africa, and Europe, which I conducted between 2013 and 2018.

Since around 2011, a recurring topic in the hacker and coworking spaces where I had been embedded since 2008 was how to "take things to the next level" and to "scale up." People were suddenly talking about makerspaces and hackerspaces as a form of early training for and experimentation with alternative career paths, outside the large corporation or well-established institution. Start-up weekends and hackathons were hosted at maker and coworking spaces, libraries, and universities. Interest increased in incubator and accelerator programs, which were understood as providing exactly that "next level" of training. These programs, largely set up in the same kinds of urban spaces as maker and hackerspaces (from garages to refurbished factory buildings), appeared to retain the kind of commitment to playful experimentation, fun and flexible work culture, peer production, and community that many had come to identify with maker and hacker culture—the promise was to connect this culture of tinkering to "the real world."

I sat down with Mike Wang for a formal interview a couple of years after he had accepted the Thiel Fellowship and asked him to reflect on the years when—for him—most of this all began:

> I was in this robotics club in high school, in Shanghai. One of the folks at XinCheJian—I think it was Paul—wrote up a blog post about a robotics competition we did. So I went onto their website [of XinCheJian], and I went "oh, wow, this is really cool! Why didn't I know about this before?" . . . I just looked up the address and went and I walked in. I think the first time I went, it was Ricky and Min Lin and David, couple other people. And then they're like, "Who are you?" I'm like, "Oh yeah, I'm just a high-school guy. I read about you guys on the Internet. Can I, like, help out or something? I don't know if there's a membership fee or something." They let me help organize the space and think things out and stuff and they're like, "As long as you don't annoy people and get some things done, you're welcome here." I was like, it sounds great. So I started out just kind of going to workshops and like . . . I basically came everyday to XinCheJian . . . And it was really good. I, like, learned a lot . . . I was there when we snaked the ethernet cable over to the neighbors to steal Internet. Took in an old, broken air con, fixed it, turned it into a beer cooler and stuff. So I really, really enjoyed . . . That summer, I came in every day. Like, I learned more that summer than like my entire school career up to that point. . . . And then, so then I went to Berkeley that summer. Every time I came back [to Shanghai] during the holidays, I would hang out.

I asked him, "When did you get the fellowship exactly?" He replied,

> I left [Berkeley] after two years to take the Thiel Fellowship. It was a big step . . . So, I mean, I heard about it my first year because a friend of mine had just gotten in the year before, and I think it was XinCheJian that had already seeded my mind with the idea there were things other than school. I really enjoyed being in Berkeley, but I think by the time I even started attending, I had this idea in my head that it's . . . I don't have to stay four years. I'll stay as long as I can get as much as I can out of classes and the school environment and if there's an opportunity to go out and start apply-ing that and try things out, not on a sort of scheduled and structured track, then I should do it, 'cause I'd already had a taste of that at XinCheJian and it was great.

What Wang describes here was more than just the curious journey of a teen-ager into young adulthood. He had learned to see his previous experiences—the robotics competitions in high school, the two years in college, his work and friendships at the Shanghai hackerspace—as resources, all of which he could flexibly draw from to position himself. These experiences made up his portfolio, a list of assets that pertained to all aspects of life.[11] Wang had transformed himself from a high school student who wasn't sure of his place at a local hackerspace into a confident young man. He had learned "self-appreciation," which the philosopher Michel Feher describes as the allocation of one's competencies and the ability to value oneself in neoliberal finance capitalism.[12]

For Wang, much of his life, from what was learned in school to intimate interactions with a friend, became assets, investments in himself. Indeed, the call for "life-long learning" (which is the standard response to fears about unemployment and job loss due to automation and AI) is a call to turn life itself into an asset that underwrites continuous self-investment. When the director of the incubator that would later invest in Wang wrote Wang's story for a news outlet, he asserted that people like Wang would "revolutionize the technology industry." This innovation revolution was ever more crucial, the director made clear in the article, during the global moment of "tech stagnation." Incubators like his own and programs like the Thiel fellowship, he argued, were a "way to spur innovation," or, as Thiel himself had put it, to "re-accelerate" the tech industry. A photo of Wang included in the article depicts him in a fashionable business suit vest with tie, the sleeves of his button-down shirt rolled up to his elbows, standing on the balcony of a high-rise building. Wang seems to be on top of the world, looking outward, past the camera, into the vast urban landscape beneath him, lit up in the dark.

By 2014, when the article about Wang came out and Thiel's book was published, the 2007–2008 financial crisis had already happened, and fears about stagnation in the tech industry and a backward educational system were intensifying. These fears were marked by a desire to regain control that many argued had been lost, taken out of people's hands by an array of forces beyond individual reach—from globalization to outsourcing and automation. The tools of the maker movement, from open source hardware platforms like the Arduino to low-cost 3D printing machines, were positioned as offering instruments and devices that would allow people to use technology to "hack" not only things but also markets, jobs, and economies. According to the standing argument, this would return control to the people. As Anderson famously argued in 2012, the 3D printer meant that "you (you!) can now set factories into motion with a mouse click."[13] Makerspaces, maker devices, and machines were seen as early spaces and tools with which to experiment with reinventing oneself, to transform the self from passive consumer to active change maker. They were pitched as the precursors of, and an evolutionary stepping-stone toward, the hardware incubators and accelerators that followed them. Makerspaces suddenly were framed as the early training programs of human capital. Bre Pettis, one of the founders of the hackerspace NYC Resistor and of MakerBot, the 3D printing company that had popularized making as a way of regaining control and individual empowerment, put it this way in a 2015 interview with *Popular*

Mechanics: "Making today is two clicks away from becoming an entrepreneur. It's about taking the steps from self-reliance to self-employment."[14]

After the financial crisis, the tech industry was shaken by an increasing awareness that the promises of the 1990s—from ideals of the information society to the dot-com boom, when economic development had appeared to be built into the Internet itself—had been broken. In response to rising criticism, however, the tech world did not take issue with the promise of technological progress itself. They doubled down, arguing that technological promise had simply not been "evenly distributed." I heard time and again from prominent figures in the tech industry that the answer was to democratize the promise of tech innovation, to intervene in and re-accelerate the tech industry. It was time, as Thiel had described it, to expand "future making" beyond elite colleges and regions like Silicon Valley. The excitement stirred by making, with its socialist pitch of returning control and empowerment to the people, was ideal for the tech industry as it attempted to re-legitimate itself.

The dot-com era had produced an ideology that normalized risk-taking.[15] This ideology was not produced in a vacuum; it emerged out of the specific historical condition of economic uncertainty and structural adjustments that characterized the 1990s. Through the "discursive production of the dot com era," Neff argues, people began to believe that they would profit from economic uncertainty and that they should embrace rather than fear the corporate changes under way: "the lure of risk—the potential for payout—adds an element of choice," Neff writes. "People are choosing to accept risk rather than merely accepting the consequences of economic structural change. Framing economic risks as desirable is one of the enduring social consequences of the dot-com era."[16] But the story of risk-taking as empowering was much harder to sustain in the years following the financial crisis, which fueled the spread of new conditions of precarious work and labor exploitation.[17] At this point, the discourse shifted from a celebratory story of risk-taking to a story of "democratization"—in other words, an attempt to expand the pool of people who considered themselves empowered to take risks. The structural shifts in the tech industry following the financial crisis were presented as an opportunity space, creating an affect of regaining control. Maker advocates began discussing the slowdown of Moore's Law, which said that transistors shrank at such a speed that every two years, twice as many transistors could fit onto a chip. This law was the engine of tech innovation and the foundation of the "old" tech corporation's business model. But although Moore's Law appeared to be faltering, the hacker Bunnie Huang famously argued that this was not a reason to despair (see chapter 3 for details); on the contrary, it was a moment of unprecedented opportunity, for the kind of innovation long reserved for high-resource tech corporations and research laboratories was now becoming available to everyone! As Moore's Law slowed, hardware costs decreased, for profits could no longer accrue through the shrinking of components and devices alone. Industrial production itself was opening up to future making, to the investment in promises.

The Socialist Pitch in Capital Investment and Higher Education

It was November 19, 2012. I returned home late from a busy evening at the Shanghai hackerspace XinCheJian. That year, making appeared to be at its peak, and China was where it all seemed to be happening. On that particular day, Chris Anderson had been invited to Beijing to speak at the NetEase Mobile Media Summit, which was sponsored by the Chinese government. But really, he tweeted, he was there for the "launch of makers in China." Anderson had quit his job as editor-in-chief of *Wired* magazine to start an open source robotics company, and this, along with his decision to speak at the summit, was the topic of conversation that day at the hackerspace. Mario, a relatively new member of XinCheJian, had likewise quit his own job in international trade, and that day he had given a promotional talk for a workshop he was planning to host at the hackerspace. The workshop was on how to "build a 3D printer in 2 days." "This is normally not achievable," he had explained, "but I will source all the parts and help you."

In an interview I conducted with him later, Mario told me that he was able to dedicate his time fully to the future of open source 3D printing because of his past life working in the oil industry. This job had provided him with the financial means to support himself and his family, whom he had moved from abroad to China. Most other makers I met could draw on neither the kind of independent wealth that Mario had nor on the reputation that Anderson had garnered—the reputation that drew international attention to his move into open source hardware. My ethnographic fieldnotes from that day (and the months before and after it) are filled with personal stories from people who had put tremendous amounts of free labor into starting and running hacker- and makerspaces. For many, this was a side job. In the evenings and on the weekends, these people hosted educational workshops, start-up weekends, BarCamps, and hackathons; they recruited new members, attended conferences, gave talks, and raised money. My notes document stories about tired bodies, jokes about bare survival, worries about the economic slowdown and job precarity, and hopes that all of this would "pay off"—there was so much *potential*, they said. The future was the maker.

China's maker advocates had just hosted the 2012 Maker Carnival, the first of a series of large China-specific maker events. The event had been a major success, attracting international attendees, including well-known American hackers, and the attention and approval of the local Chinese government (see chapter 2 for details). The fact that one of the main organizers had put up significant personal funds to make the event happen did not cloud the excitement. During this period, the number of maker- and hackerspaces in China was increasing exponentially (as they were in other countries). At the Shanghai hackerspace, there were frequent visitors from the upper management of major foreign tech companies (from Intel to Twitter) and reporters from foreign news media. Everyone at the Shanghai hackerspace was convinced: the exhaustion, the endless hours of free labor, the investment of passion, it was all worth it. The growing international attention suggested that there was something unique about making in China. Intel and other major Western tech companies had begun sponsoring Maker Faires and competitions in China; China seemed the place to be. I too felt this excitement. Previously, my research into making had often received puzzled looks, and when I explained that I focused on making in China, I got even more puzzled looks. People asked what my research on making had to do with the fields of information, communication, and digital technology. People told me that China couldn't possibly have innovation. People accused me of applying a Western framework to China. Therefore, when I began being invited to give talks on the topic, to write opinion pieces, and was interviewed by journalists, I was surprised; the increase in attention and the shift in attitude were remarkable.

In Shanghai, conversation returned again and again to what was happening in Shenzhen. Shenzhen was a place where they don't even need makerspaces, I was often told: "It's a city-size tech-shop, a city-size hackerspace," a "candyland for electrical engineers and mechanical engineers." There was talk of a Western venture capital firm that had set up a hardware incubator program in Shenzhen, partnering with the Chinese open source hardware company Seeed Studio, which had made a name for itself in international maker and tech networks (see chapter 2). This partnership was about "taking open source hardware to the next level." They chose Shenzhen, I was told, because it was becoming known as a place where start-ups could prototype the next stage of technology innovation. The vision that China's maker advocates had been articulating since 2008 was seemingly becoming a reality: China was becoming a key site for democratizing making, in the process shifting from a producer of cheap, low-quality copies to a global innovation hub (see chapter 3). The new Shenzhen-based hardware incubator struck many as more than just your average start-up training program, for it promised to build on and "accelerate" the maker ideals: community formation, cooperative sharing, peer production, and returning ownership of the means of production to the people.

The incubator's next cohort (its second) would be comprised of eleven foreign (mostly American) and two Chinese start-ups (allegedly out of hundreds of applicants). By the end of the three-month program, the start-ups had run successful crowdfunding campaigns via Kickstarter and Indiegogo.

American media covered their product pitches in San Francisco, the culmination of the program, with much excitement, claiming that these start-ups "were building on the maker movement" and that they were "lean" and "agile," just like software start-ups.[18] The program was "legit," many told me, because it extended and enabled the ideals of the maker movement, and because it was modeled on familiar approaches to starting up a tech business. Its visibility in Silicon Valley suggested it had the "right" connections to venture capital and tech innovation, and its reputation for working with "ethical" factories in Shenzhen gave it respectability.

Serena and Antoine, two friends from graduate school who had started their own design consultancy after graduation, began taking an interest in the incubator program. Since grad school, they had been presenting their explorative and artistic open source hardware projects at Maker Faires. They were working on a low-cost lighting system for a slum community in Southeast Asia that had been devastated by a series of fires. The goal was to build an affordable kit of simple and safe electronics that would provide lighting to these communities, which were largely cut off from basic infrastructure, including stable electricity. The prototypes of the kits were beautiful laser-cut pieces that built on the target community's aesthetics of craft. When Serena and Antoine presented these kits at Maker Faires, they garnered significant interest among makers, families, and educators. It seemed like an ideal project to pitch to the hardware incubator, which promised to help makers scale their inventions beyond the world of already fairly empowered geeks; if it could be mass-produced, the lighting kit could make a real difference for low-income communities across the globe who could not afford basic electronics infrastructure.

The three of us plotted how to make our way to Shenzhen, Serena and Antoine as a start-up, and me as their on-site ethnographer. Serena and Antoine submitted a short video to "pitch" their team, a key requirement for admission to the incubator program. A couple of weeks later, I got an email from Serena: "Dude, we got in! We are going to Shenzhen! Let's talk soon to figure out shared housing stuff." Things moved quickly. Two months later, Serena, Antoine, and I sat in the offices of the hardware incubator, listening to an introduction to the program from Evan (the director) and Mike (the program manager). "This is where shit is happening," Mike proclaimed, bouncing up and down in the front of the room, pointing to the city landscape that spread out ten floors below us—the electronics markets of Huaqiangbei (华强北). The first week of the program, the director said, was an accelerated version of the 111-days-long incubator program; in that first week, we had to learn how to source from and take inspiration from Huaqiangbei in order to build a working prototype within a week. We would build nothing from scratch; we would instead identify platforms and products that already existed in the markets and hack them, opening them up and turning them into something else. Each start-up was separated from its team and re-assigned to temporary teams in order "to get to know each other's strengths" and "get to know Shenzhen speed." As Evan reminded us, Shenzhen meant speed, and Shenzhen speed was one of the key advantages this program had over typical incubators based in the Bay Area. "Take this seriously," he said. "Be flexible. Take your time [in the markets], ask questions, explore, observe and power up! You don't have much time."

The program's first week shaped in important ways how we would all relate to one another. The program's success, and the success of the ventures it invested in, hinged on how well each of us managed to mobilize our individual capacities to make each team unique. "This is your key asset," explained Peter O'Melly, the program's co-founder (who had also founded the venture capital firm that had funded the incubator). The first week seemed like a test of sorts, a trial run to assess whether you were truly cut out for this, whether you would find your place on your start-up team and in the program. I felt immense pressure and even desire to prove my value and fit for the program. On the first day, the director took me aside and explained that he had been thinking hard about "how to make your participation fit." He said with utmost seriousness, "Timing is everything for us and the start-ups," and explained that "we can't have anything or anybody here that distracts them . . . this is like a private club and we can't have any outsiders here." The next day I found myself explaining over email the "value" of ethnographic research and how it differed from a full-on "outsider position," and after some back-and-forth, they informed me that the director and program manager had decided they could make my participation work, if I became a member of one of the start-ups. The program manager asked, "Isn't anyway the point of your research to do exactly what the start-up is doing? Becoming a start-up seems like the ultimate way to show the value of your research!" I agreed to the arrangement under the condition that I would be open about my positionality as ethnographer with the start-up teams and the other people I'd engage with in the program. In these exchanges, I realized, I had positioned myself based on the value I'd add, the unique portfolio that an ethnographer would provide; I was allocating and marketing my competencies, not unlike Wang had done, with whom I began this chapter.

The incubator's demand to self-identify through a language of investment was perhaps so easy to fulfill exactly because it had become so pervasive at the university itself. During this period, I had been on the academic job market, submitting job applications to universities in China, Europe, and the United States and traveling for the occasional job interview. Many of the people I met through my research were impressed by but also doubtful of a career in academia. Universities were perceived as symbolizing the docile and heavy bureaucracy that many of these people had set out to disrupt, to prototype alternatives to. Universities were seen as heavy, slow, incapable of addressing the societal and economic challenges of the twenty-first century.[19] But this view was not common only among the maker, open source, and tech start-up crowds; I heard the same rhetoric from inside academia itself. Calls to make education "lean" and to encourage faculty and students to become entrepreneurial and self-invested were pervasive among university administrators, educators, policy makers, and even some scholars in my fields. Academia has been plagued by the dismantling of tenure-track jobs, the increase in adjunct positions, and the reduction of faculty governance.[20] When I was on the job market, I felt that there was little being done, either in the academy or beyond it, to counteract these trends. There seemed to be no alternative to the "neoliberal privatization" of the university, given the universality of the "insistence that even publicly funded universities" should operate like businesses and "embrace a lot of the same practices and value systems as for-profit corporations."[21] It was this rhetoric of inevitability—the sense that there was no way to make education relevant except by making it lean, by training students and faculty in entrepreneurial "agility"—that legitimized universities' increasing reliance on short-term and precarious contract labor in the academy, their rush to adopt MOOCs (Massive Open Online Courses), and their investment in university incubator and entrepreneurship programs.

The people admitted to the Shenzhen incubator program seemed to be ahead of the curve, and they saw academia as trying to catch up with the latest trends in industry. I witnessed these catch-up attempts myself; just two months before moving to Shenzhen, I participated in a workshop for early career scholars of PhD students and postdoctoral fellows, hosted by senior colleagues in STS (science and technology studies). Participants had to form teams of four to five people, come up with a product idea, and "pitch" the idea at the end of the four-day workshop. The task was to learn how to identify the key "value" of our scholarship in order to help us better position ourselves in the competitive academic job market. "You have to learn how to identify your own unique selling point," one of the senior scholars—himself caught in a non-tenured and precarious research position—told me in between sessions. "I wish I had known better when I was on the job market." Over the four days, we listened to lectures on Latour and Foucault, conducted quick and dirty consumer validation research, prototyped a product, and pitched our "intervention." This particular workshop was one of many similar efforts that I would encounter at the university in the years to follow, what Gary Hall describes as the "uberfication of the university":[22] a process of economization that is seductive because it is masked behind a hopeful rhetoric of prototyping "alternatives" for broken educational systems, cooperative sharing, and capacity building.

It is exactly this promise that I call the "socialist pitch"—the pervasive promise of hopeful intervention—that capitalizes on people's desires for empowerment and yearning for institutional change. It was the socialist pitch that fueled and enabled programs like the Shenzhen-based incubator. The socialist pitch encouraged us to invest in ourselves and in a seemingly hopeful project to prototype the future of work, of education, and societal change; it framed the start-up teams as forerunners, as shaping and changing both the conversation and institutional structures. The socialist pitch is a powerful rhetoric of *inevitability* (there is no other way forward) and *anticipation* (the promise of a better future, rendered palpable in the present).[23] In neoliberal capitalism, Murphy argues, quantitative measures are enriched with affect. Numbers like GDP growth and measures such as the correlation between education of young women and their ability to produce future economic actors "lure feeling" and "propagate imaginaries" of national development and a better world. In the neoliberal economists' framing of human capital, bodies become a site for an "anticipatory, future-oriented calculation of value."[24] The socialist pitch, so pervasive in the tech industry and tech research communities like human-computer interaction, draws on the affective bond between long-held desires for societal and economic change and the *promise* that technology production will be democratized, and people thus empowered to intervene *at scale*. At the incubator, people *felt* they were "hacking" economies of scale, global supply chains, and industrial production.

There is a risk to rendering the participants in these programs as dupes, as blinded by promises that are then unmasked by the critical scholar as tools of inequality, precarity, and exploitation.[25] This is not my intention here. One of my goals is to show how implicated and complicit the very institutions of scholarly practice are in proliferating and anticipating this mindset of inevitability and the promise of interventionist agency. The condition of the socialist pitch has been co-produced by universities and a variety of tech research institutes,[26] as they render not only inevitable but desirable and hopeful online degrees, makerspaces, incubators, the quantification of evaluation, and the adoption of business methods from the lean start-up to design thinking. These are framed as the only way to make education relevant once again, able to make an intervention. These efforts do not build better public education. They simply imbue the neoliberal demand that individuals invest in their selves with affect, with hope, with promise, with future making.

Market Devices of Finance Capitalism

On the second day of the program, Peter O'Melly, the CEO of the Western venture capital firm that had funded the Shenzhen incubator, gave a speech that reminded me of the kind of motivational, often deliberately personal talks that have become notorious in the tech industry and through organizations like TED. O'Melly was giving his famous "rocket-fuel speech," Evan, the program director, explained to us. The speech, meant to motivate the start-up teams for the weeks to come, also modeled the kind of subjectivity that promised success—confident, survivalist, optimist, interventionist. This is how it began:

> I am here to talk about the meaning of life, Maslow's hierarchy of needs. You work up from very basic physiological needs like food and sex, and then work up higher the chain of existence. As entrepreneurs, we are high up at this chain, because we self-actualize, even if we don't get sex [laughs]. We live in a very unique time of this planet, where makers have the opportunity to affect billions of people. Charles Darwin once said, "it is not the strongest of the species that survives nor the most intelligent, but the most responsive to change." We live in an environment of change, and it is so important to accept this message. People who do this best are those who listen to their customers, mentors and to themselves and make that change. Through entrepreneurship, we have disproportionate power to affect the world.

O'Melly is a charismatic man. He embodies the kind of exuberant, self-invested geek-turned-entrepreneur he promised all of us in the room had the potential to become. Being an entrepreneur was a position of power, he told us that morning, a final step in the evolutionary development of the human species, the species that he said was "the strongest" because it was "the most responsive to change." According to O'Melly, the entrepreneur must not only adapt to change but also "make change." In O'Melly's vision, the entrepreneur holds responsibility; he is a change maker urgently needed during society's "moment of drastic change" and "pressure to move quickly." And he reminded us that social disruption and change were reasons to celebrate, not to despair: "difficult times are an encouragement to move faster. If you stay still, you will be dead!" The entrepreneur must "see opportunity where others get stuck" in order to survive and even thrive during the coming upheaval.

"You, the people admitted to our program," he explained, are "the forerunners," people who will shape and guide this societal and economic transformation. We would develop devices, tools, machines, and platforms that enable others to see promise and even pleasure in hacking—technology, societies, markets, economies. In an interview with me, Evan had told me that those selected to join the program displayed both a strong affinity for technological mastery and a readiness to extend the fun and pleasure of that mastery into the realm of markets, education, work, industry, and economies of scale. He explained that the program enabled a transformation from geek into entrepreneur: "Not all makers are made to be entrepreneurs," he told me, "not all of them want to be entrepreneurs. I'd say a very small minority of them want to be entrepreneurs. For the most part, they get there because they want to have fun. Then there is the slide, or the switch, to become an entrepreneur. This is about being able to still have fun with what you are doing [entrepreneurship]."

Crucial to this "slide," to making the "switch" into entrepreneurship, was the capacity to approach this new "world" of entrepreneurship with the same enthusiasm, the same attitude of "having fun" that makers applied to tinkering with technology itself. It was about learning how to derive pleasure not only from hacking machines but from hacking markets.

Making this shift also requires making a shift in how the subjectivity of the geek and the hacker is construed. As STS scholar Nathan Ensmenger shows, the pervasive cultural imaginary of the programmer as predominantly inter-ested in technological questions, removed from reality, was not a natural or inherent quality to the "geek" or "nerd" but the product of social construction during the 1960s software crisis.[27] This was the moment in which management, especially middle management, began to decry the work of programmers as a craft-like enterprise that lacked structure and a guarantee of economic profit-ability.[28] At first, programmers seemed able to resist the Taylorization of the workforce, in large part because they attempted to legitimize their work by ways of "using masculinity as a cultural resource that aspiring professionals could draw upon in order to improve the standing of their field."[29] Fears over the broadening precarization of tech and creative work fueled the demand placed on geeks to counteract the feminization of their labor and to re-invent themselves as entrepreneurs—in Evan's words: "This is just the way the world is going at the moment; with more automation, and these shifts in the tech industry, all of those people will have to create their own jobs now." If geeks managed to become "change makers" (the term media and investors often used to describe the participants in the hardware incubator), then they were empowered to hack machines and also economic and societal realities.

To become a change maker, we had to identify our unique assets, includ-ing the human capital on our teams: "One of your key assets is your team. The team is everything," O'Melly told the start-ups. "With the right team, you will get people to care about what you do. You get them to understand quickly." To identify our individual assets, each of us had to take a personality test at the end of the first week. The personality test was designed by Adam, an employee at the venture capital firm: "Adam is our man. He invented a method called 'Wholebrain Thinking,'" Evan told the start-ups. As Adam himself put it, the method was a "a rescue package [that] involves combining science and psy-chology . . . to bring a brighter future." "What is the 'so what' here?" he asked the start-ups rhetorically, "So?! Life expectancy of Fortune 500 companies has fallen from 50 years to around 10 years in just a decade. The world is hungry and bankrupt." The "wholebrain thinking" method was an instrument to help entrepreneurs intervene in this moment of the vanishing large corporation,[30] which marked drastic shifts in the industry, by making the entrepreneurs iden-tify and better leverage their own abilities and strengths. The basic under-pinnings of the personality test were developed by Ned Herrmann when he was leading an educational program for managers at GE. The method rests on the idea that each individual's capacities can be quantified into a unique spread of four dominant modes of thinking: analytical, sequential, interpersonal, and imaginative. Each successful start-up, Adam explained, was comprised of a team that represented all four dominant modes of thinking. "The key is," he explained, "to recruit a wholebrain team."

This personality test was one of a series of "market devices" deployed by the incubator. These market devices are instruments and methods that render things attractive to the logics of finance capital—they assign them an affect of promise and anticipation; they have become common in tech entrepreneurship and design consulting. As STS scholarship on economization defines them, market devices are translation machines, turning one set of relations into another that serves the market; they produce what STS scholar Michel Callon calls "agencement,"[31] the rendering of things, behaviors, and processes as economic. This process of economization is what allows an institution, an action, an actor, or an object to be considered as being economic.[32] As STS scholars have shown, the equipment of economics (such as market devices) matters, for economic models and techniques of abstraction not only analyze markets but actually co-produce them. In the same way, economists do not merely study markets; they bring markets into being.[33] Economic actors are thus agglomerations and combinations of human beings, material objects, technical systems, texts, algorithms, and instruments.

I argue that market devices, and the agencement they produce, take on a particular shape in contemporary neoliberal finance capitalism. Their primary aim is to render the self (and life itself) economic in a particular way: as attractive to finance speculation. Financial markets are operated and dominated by large banks and institutional investors, and as sociologist Jerry Davis shows, the traditional industrial corporation must now learn to reposition itself as attractive to these investment institutions. The priority of these corporations— and of the state itself—is to sustain the confidence of investors. Employers and governments are no longer in a "position to promise lifelong careers to their employees and constituents," the philosopher Michel Feher argues. Instead, "it is now up to the job applicants to make themselves valuable . . . their ability to find work depends more on the credit attributed to their human capital than on collective agreements about salaries and labor conditions."[34] In other words, workers must make themselves attractive to investors—sometimes to employers who invest in them, sometimes to direct investors like venture capitalists. Devices like the Herrmann Test were meant to guide us to think about ourselves in terms of self-worth and human capital. After we all took the Herrmann test, Adam printed our results and hung them on a wall next to the entrance to the program's office space. "This will remind you," the program director explained, "to always think of where your strengths are. You can use this to improve yourself. Behaviors can change. All of this here [the incubator] is really a learning experience. It's all about self-awareness and consciousness, about knowing what you do well."

The Herrmann test was one of a series of market devices of finance capital that the incubator used to teach people a particular kind of self-awareness and self-consciousness: to teach us to re-evaluate ourselves through the logics of accreditation, the valuation of capital—specifically, through the valuation of *human* capital.[35] Together, these market devices formed a toolkit to render oneself attractive to the logics of investment. They included lean start-up methods, the business canvas model, design thinking, and human-centered design methods. Methods like these have become pervasive not only in incubators but in universities, corporations, management, and even governments. They are aimed at training people to approach the world through a logic of "rational and scientific calculation" that renders complex societal and economic challenges as "solvable" by technological intervention[36]—they work by inducing a feeling of change and hope. These market devices come with recipes and techniques that promise universal applicability, summarized in downloadable PDFs and colorful brochures that are readily available to both teachers and students. How-to guides and materials showing these methods in action depict a diverse (often young) array of people in front of colorful Post-It walls, interacting with children in urban slum communities and prototyping altruistic designs with the people of an African village. This aesthetic of the cheerful, optimistic promise of intervention, of making a difference, of making the world a better place by bringing justice to the people and to places that have been exploited—this is the technique of the socialist pitch.

The key impact, however, is seldom systemic change in communities selected as sites of intervention.[37] The often short-term engagements instead function as powerful portfolio pieces for the interventionists themselves; they are assets, for the creative entrepreneur and designer, who is trained to invest in and appreciate her own human capital. These market devices of finance capital, couched as helping scale "open innovation," "participatory design," and "peer production," enable contemporary neoliberal finance capitalism to do away with traditional wage labor-capital relations and to demand of workers a new economic subjectivity of self-investment.[38]

Incubators teach people how to retrain themselves, to remake themselves as attractive to the anticipatory logics of finance capital. As O'Melly put it during his rocket-fuel speech, "Design for the future, not for the now!" It is this anticipatory logic that renders the uncertainty and complexity of contemporary economic and societal shifts as under control, as manageable and indeed "hackable" by the geek-turned-entrepreneur.[39] Market devices, from the lean start-up and open innovation to human-centered design, are the instruments that entice people to produce themselves as human capital. The fact that these methods were introduced or written by people in powerful positions in the tech and creative industries and at prestigious universities gave them legitimacy and imbued them with hopeful promise.[40] In other words, these devices themselves are forms of the socialist pitch; they seem to re-enable community where there was none, to empower people who have lost control, to produce national development and a better world in response to a moment perceived as drastic upheaval—upheaval that is in fact caused by the "opening" and democratization of innovation itself.[41] These market devices of neoliberal finance capitalism, which are central to producing the anticipatory economy of future making, are fueled by yearnings for alternatives to contemporary structures. They enabled the behavioral shift that Feher shows financial capital demands, from socialist commitments of hacking, community, and solidarity to "what is doable"—action over thinking.

Many of the start-up teams that joined the Shenzhen incubator had been committed to systemic change, including projects that ranged from addressing infrastructural lack in urban slum communities to democratizing open source hardware tinkering beyond an elite of affluent makers. The socialist pitch of incubation, of human-centered design, and lean start-up methods created an affective bond between these ideals and the interests of finance capital and market speculation; it rendered the cultivation of human capital as both inevitable and hopeful. But this affect was designed to attract investment, to make the self a commodity to be invested in. Over the duration of the program, I observed how many of the ideals that had motivated people to join the program were mobilized and rewritten to attract investment. The program trained the teams to attract the interests of two sets of audiences: finance capital investors and a potential consumer base, which would be reached via crowdfunding platforms. While considerable time and labor went into learning how to reach both potential investors and potential customers, the customers were secondary. The successful crowdfunding campaigns that the start-ups produced by the end of the program were not designed to sell products—indeed, most hardware start-ups barely break even or even lose money from the sales made through their crowdfunding campaigns.[42] Instead, successful crowdfunding campaigns were used to show that the product had a potential market. Investors look for these indicators to minimize their risks in investing. A successful Kickstarter or Indiegogo campaign demonstrates that the idea and the start-up have promise.

"Backers" of crowdfunding campaigns do not buy only a product (although they might eventually receive one in the mail); they "invest" in the story of a product, a team's pitch, the credibility of the start-up. The key task of the three-month incubator program in Shenzhen was to produce the types of personal accreditation, the performance of affect, that would attract investors. The hardware prototypes produced by the start-ups as they worked with prototyping facilities, factories, and suppliers in Shenzhen, the marketing of the prototype through crowdfunding, and the final pitch in front of investors (which followed a particular aesthetic and mode of storytelling) were all aimed at this accreditation. The incubator's primary training objective was to teach start-ups how to tell the story right, through both the material (the prototype) and the discursive (the Kickstarter video).

Each week for the three months, the incubator hosted a test run of the start-ups' final pitch, with feedback from the fellow start-up teams and the program's director and manager. Mentors—at the time exclusively male—came to the office to give talks on the lay of the land of investment, to provide insights into hardware prototyping, and to share strategies for working with factories and suppliers. Over time, these sessions and interactions shaped not only the teams' pitches but also the ways they positioned their commitments. The start-ups' original ideals—to support slum communities, to provide science education for poor children, to enable subversive street play—were re-scripted so that they fueled a new "maker marketplace." The products that were pitched to investors and media in San Francisco at the end of the program had been transformed into cutting-edge IoT platforms and open source hardware kits— yearnings for justice and social impact were "upgraded" into sexy products that made tangible the promise of making. In other words, these products embodied the idea that by learning how to make *things*, people were investing in their own capacity to remake *themselves*, to prototype themselves as "change makers." What was sold that "demo day" in San Francisco were not products or maker kits but the promise of future gains, of enhancements to investors' portfolios, of contributions to the next share sale. The start-ups were taught to produce not products to be sold, but promises to be invested in.

This created a particular kind of affective relationship between business and customers: like start-ups, customers of their smart hardware products were also positioned to invest in transforming themselves. "We don't treat them as customers," Serena told me while we practiced for another pitch session. "We treat them as our partners. These guys will help us seduce other people into making a habit of tinkering with their device, of developing an attitude of hacking. So the product is seducing a novice into becoming an expert [in hacking]. Transforming somebody who has an inkling, who might be eager to change . . . this is a huge market!" The IoT platforms and smart hardware kits the start-ups pitched promised to "seduce" customers into seeing themselves as aspiring change makers. The point was not to make a profit from selling hardware devices, but to entice others to invest in the promise of tinkering, hacking, and open source hardware in order to regain control, bring power to the people, and participate in industrial production.

The program director often described this socialist pitch for behavioral change as a way of training people how "to let go of tech" and "push . . . the teams out into the real world." By the end of the program, participants had internalized the view that "hacking" life was key to one's success. As one of the participants put it when I interviewed him after the program:

> For me, [the incubator][43] was really eye-opening in terms of how things work in the real world, outside of college. I was still in college at that time [when I joined the incubator]. I lived in this very idealistic world at that time . . . For me, [the incubator] was a big learning experience . . . our team failed in the end, not because we lacked passion for technology, but because we were acting like naïve five-year-olds, idealistic and excited about something new and shiny. The teams that succeeded were the ones that really thought about economics. It's a reality check. If you want to be in the hardware world, you need to earn money, you need to make a product. That's the goal. I don't think we were thinking about that enough. That's why our team failed.

He attributes personal success or failure not to the volatility of a market that runs on speculation but to the ability to "think economically," to transform the self from an idealistic "five-year-old" into a savvy self-economizing actor. It is this transformation from the "maker," who finds pleasure in hacking machines, to the "change maker," who finds pleasure in hacking her own human capital, that the hardware incubator provided training for and in turn invested in. The incubator did not invest in hardware but in the promise of the maker movement. It provided accreditation for the story of a future grounded in open source hardware, consumer empowerment, and democratization of industrial production. For investors, it does not matter if the promises of the maker movement ever materialize. It does not matter if industrial production will be democratized. What matters is that people continue to "buy into" this promise. The early machines and devices that were successful, from the 3D printing company MakerBot to the Pebble Smart Watch, all sold this promise. These companies and their founders modeled a seemingly inevitable evolutionary path of human development, from maker to tech entrepreneur, from geek to economic actor—from, as MakerBot co-founder Bre Pettis put it, being "self-reliant" to "self-employed." Indeed, MakerBot's socialist pitch was so successful that the company was sold to the 3D printing firm Stratasys for more than USD 600 million—a sale that was made possible by the closing of large chunks of MakerBot's open source repository. MakerBot has since been denounced by maker advocates and academics alike as selling out the values and commitments of open source hardware, as exemplifying how a once authentic counterculture can be co-opted by capital's endless reach.[44] The all too quick conclusion that the maker movement was a failure, that it too was subsumed by capitalism, however, covers up the specific processes of economization I have described so far. Both the story of success and the story of failure are crucial market devices of finance capitalism; while success stories produce feelings of promise and progress (and thus lure investment), stories of failure enable investors to reduce risks and legitimize the redirecting of investment to the next success story, the next promising start-up, the next "big idea." The story of a co-opted hacker movement (via making or via investment) further naturalizes the idea that there exists first an innocent countercultural subject position (the hacktivist) that eventually (and inevitably) will be torqued to produce economic value. What if we began instead from the understanding that there are no innocent and ideal positions, and that what is available to us is always already partially compromised? We might, then, be able to step sideways, and out of the seemingly endless cycle of success and failure, resistance and co-optation (a thread I continue in chapter 5).

Shenzhen Speed

The final "demo day" at Autodesk in San Francisco brought together an intimate gathering of investors, media, and other "friends" from the tech industry. When Evan took to the stage to introduce the thirteen start-ups, he too gave a pitch: "China is not any longer just about outsourcing. It's a lot more than that," he began. On the screen behind him appeared the words "knowledge base," "leveraging expertise," "integrated ecosystem," "speed," "quality," "price," "partner." He went on: "It's about leveraging the knowledge base, it's about an integrated ecosystem, really, which are now a completely worldwide phenomenon. It's about using speed, quality, and price to get to the point where all those hardware startups have only one goal during the first three months: to build a manufacture-able product. Everything that you see today started from almost zero, and why it matters is because the only benefit that a startup has is time, and so getting to the market as quick as possible with a product that can sell is critical."

The central principle of neoliberal finance capitalism is to make not only the self, but also regions desirable to investors. As Feher writes, "states rule through the attractiveness of a territory . . . the goal is to make the territory under your jurisdiction as attractive as possible to international investors."[45] At the hardware incubator, Shenzhen (and China writ large) was promoted as an asset; the incubator here directly appropriated the "invention of Shenzhen" as I have documented it in chapter 3: China's "otherness" (its fearless experimenter mentality, its backwardness and lax regulations, its business mentality) was translated into a unique opportunity space for investment. Shenzhen has thus come to operate like a phantasmagram—as the historian Michelle Murphy defines the term; a region articulated as a "massive material-semiotic-affective-infrastructural presence."[46] Its history of economization and of the Special Economic Zone were enriched with the affect of interventionist capacity. As Evan reminded the crowd that had gathered to hear the start-ups' pitches at

FIGURE 4.1. "Shenzhen speed"—many told me—was felt in Huaqiangbei in particular, where the noise of taping and packaging filled the afternoon and evening soundscape and its streets were buzzing with the electric bikes of "kuaidi" (fast delivery service)—Huaqiangbei "demo-ed" Shenzhen speed in practice. Photograph by the author.

Autodesk, Shenzhen stood for so much more than cheap labor. Most crucially, it embodied speed, specifically Shenzhen speed. As many people described it to me, "Speed in China is sudu. Very fast speed is Shenzhen sudu."

Start-ups had to learn speed, which was framed as their key advantage over the large corporation. By "accelerating" the time it takes to bring a new product to market, one increases one's attractiveness to investors. The start-up can only compete with traditional "economies of scale" by stepping outside their dependency on product sales and attracting finance capital. Shenzhen was enriched with this affect of speed—it promised to teach people how to accelerate.

"Our age is obsessed with speed," argues feminist STS scholar Judy Wajc-man. "Faster cars, faster trains, faster broadband, even speed dating. Speed is sexy, and digital devices are constantly sold to us as efficient, time-saving tools that promote an exciting, action-packed lifestyle . . . the typical mantra of CEOs is that technology is pushing us faster and so we have to adapt to new ways of doing business in a world of screens, texts, cell phones, information all over you."[47] As Wajcman notes, the contemporary allure of speed dates back to nineteenth-century narratives of modern technological progress that contrasted the city, an image of speed and modernity, to the slow pace of country life. Speed has become understood as inextricably linked to modern technological progress and capitalism: "time becomes beyond control as distance disappears in a world of instantaneous and simultaneous events."[48]

As time continues to speed up, it seems to be careering out of our control. Shenzhen was positioned as the ultimate laboratory for learning how to regain a sense of control over time; control would be regained by internalizing Shenzhen speed. When I asked what key challenges the incubator confronted, Evan explained to me that

> The hardest part is really to make these people entrepreneurs. They come from the geeky side, and now they have to learn how to communicate, how to collaborate, and how to work with a partner. Manufacturing here [in Shenzhen] demands exactly that . . . you either go there or you call them, but either way, you make shit happen. Communication is key. It's a people thing! If they can overcome that [their geeky side of lacking communicative skills], then they are a big step ahead, because the more you invest in a relationship, the better it gets, the more trust is there. It becomes very beneficial for everyone behind it. After that, it's a matter of smarts, street smarts. About looking at what's out there, being able to combine things, being able to be fast. The only advantage they have is speed—the time it takes to bring it to the market. Being fast, smart, and moving forward every day is key.

The start-ups needed to learn to turn speed into an asset—and embedding themselves in Shenzhen's manufacturing "ecosystem" promised them to do so. But how was Shenzhen rendered attractive to the start-ups? Using the same techniques that the start-ups themselves were learning to master. Shenzhen was articulated as offering a unique "portfolio" of economic development: its histories of SEZ speed, rapid growth, and economies of scale were articulated as unique assets. Its enduring "backwardness"—its copycat culture, its lax regulations, and capacities to economize rather than innovate—were presented as an opportunity space for the (Western) maker and hacker who hoped to become an agent of change.

This particular affect of hopeful intervention attached to the region had to be actively cultivated, then and now. The "invention of Shenzhen" as an innovation powerhouse is on the backs of migrants' labor, dreams and desires for a better life. Take once more the notion of "Shenzhen speed"; the phrase was coined during the reforms that accompanied China's economic opening in the 1980s—during a time of "testing out how far you could walk away from proscribed limits."[49] The reforms were a product of a shifting ideology in China. After Mao Zedong's death, Deng Xiaoping argued that China's future rested on a shift from socialist to capitalist principles of economic development. To appease those who followed a Mao Zedong ideology, he declared four Special Economic Zones to test his thesis. Shenzhen was one of them, and, because of its physical proximity to Hong Kong, it was considered the key laboratory for China's transformation of socialist ideals into capitalist market logics, of revolutionary subjects into entrepreneurial agents. "Shenzhen speed," which ostensibly stood for experiments in both economic and political liberalization, eventually became depoliticized. As anthropologist Mary Ann O'Donnell reminds us, the outcome of Tiananmen in 1989 firmly established that while China's political leadership granted economic agency, it continued to tightly control political agency. The message was, she writes, that "economic reform was not only making China rich and powerful, but was also [and only] doing so under the leadership of the CCP."[50]

Shenzhen speed thus stands for "neoliberal governance that downplayed economic inequality and exploitation" by promising individual opportunity and inducing desires and dreams of future economic development. Shenzhen speed was only possible because of the cultivation of a self-invested migrant who desired a better future for herself and for the nation. This subjectivity was relentlessly pushed by the state, which unleashed pervasive media campaigns that called upon migrant workers to nurture their individualized abilities, ideals, and aspirations[51]—to invest in themselves, to become "a new human," enhancing ordinary citizens' sense of new possibilities that lay within reach. By upgrading themselves, citizens would build a "new" China that would at last attain material and moral parity with the West. People endured exploitation and vulnerabilities in the present and invested in their human capital because of their affective engagement with imagined futures. The Maoist rhetoric of self-sacrifice for the nation and collective interest was transformed into a neoliberal discourse of self-reliance, initiative, risk-taking, autonomy, and competition.[52] The historian Eric Florence shows that migrant workers from rural China were made responsible to "model" this construction of a post-Mao and postsocialist China; they bore the responsibility and precarity that came with this new lifestyle and the "Shenzhen spirit."[53] They were called upon to invest in themselves, to improve themselves in order to raise their suzhi (素质—a person's quality and capacities). They were called the "builders" of China's future, a designation that identified them with Shenzhen, but they were not granted household registration status, and they therefore lacked access to basic social benefits, health insurance, and workers' protection.

The concept of Shenzhen speed thus condenses "the precariousness and liminality that characterizes and continues to define migrant workers' conditions and the politico-institutional arrangements that have enabled the exploitative regimes of production."[54] When investors celebrated Shenzhen speed as hopeful and promising for the geek-to-become-entrepreneur, they masked

how this very self-transformation would demand a version of the precariousness and exploitation that generations of Chinese had already "modeled." Chinese migrant workers, so to speak, were made to prototype the ideological
apparatus of neoliberal finance capitalism, and the making of Shenzhen, and
with it China, as an attractive region for further investment. The celebratory
articulation of "Shenzhen speed" as deployed by technologists, politicians, and
investors today naturalizes the region as inherently and uniquely self-reliant,
resourceful, and experimental, masking the labor that was necessary to produce it as such.

Shenzhen became attractive to venture capital in the very moment that
its own machinery of "promise production" was stuttering. At the incubator,
stories of a "renewed" reality of the tech industry were a constant, told again
and again by the investors and employees tied to the program. Key to these
stories was a sharing of one's "personal journey" of risk-taking, suffering, and
transformation. "You know back then, during the dot-com era, things were
very different," one of the mentors, who described himself as an entrepreneur-
turned-VC, began. When he visited the incubator, he explained to us the history
of his early forays in Silicon Valley in the late 1990s. He asked the start-ups to sit
in a circle, with him standing in front of everyone, his loud voice penetrating the
room. His tone was conversational, but his speech permitted little interruption.
The start-ups sat quietly as he spoke:

> About 14 years ago I was self-employed writing software during the dot-
> com era. And it was a lot of fun. And I fancied myself an entrepreneur. You
> know I was getting ready to kind of—I actually sold a product that I built as
> a consultant . . . this was in like the middle of 1999. And I went around town
> in New York City firing [up] my customers. You know, like "I am gonna
> start like pets.com. I am gonna start like waterbottle.com. I am gonna get
> 30 million dollars of venture capital and go public in 6 months." 'Cause
> that's how crazy it was. It was a good time to be in technology. If I needed
> money, I'd just call up my friends. My partners were from Harvard Business
> School. So I'd call up their friends. I ask them, "so who is your next rich
> friend?" And they said, "Nick Negroponte of the MIT Media Lab." I said,
> "all right, I am going to go up." We took the Amtrak up to Boston, met Nick
> Negroponte. He showed us some pretty cool robots and other fun stuff,
> and he goes, "Here's a check for a million dollars." Boom, done. I am like,
> "Wow, this is easy. Two down, 28 million USD to go." Those times have
> passed. Today is an era of "bootstrapping," because really you don't need
> money to start a company.

Here he paused and looked around the room, as if waiting for a question.
"You bootstrap. You have to do your homework though. This means you have

to know your investor. The only resource you need is *you*! You have to retrain yourself to see market windows and opportunities." He articulated the central message of the incubator: the key investment to be made was "you"!

The prototype for these contemporary modes of self-investment was the Chinese migrant, the self-made Chinese entrepreneur, the worker who came from nothing and re-invented himself—from worker to engineer to factory owner. This figure of the Chinese self-made entrepreneur was the moral prototype for the subjectivity of self-investment that investors demanded of hardware start-ups. I observed how Chinese men who had succeeded in their self-upgrade from migrant to entrepreneur were celebrated as "awesome partners," as people "who blow your mind" and who "are eager to help," who were "humble" and "down-to-earth." These stories of success were crucial in creating an affective affinity and shared sensibilities between the start-up teams and the factories they worked with. In early 2013, Zach Hoeken Smith, co-founder of MakerBot, who had spent much time in Shenzhen on behalf of his company, published one such story on his blog. The post was titled, "Meet a Shenzhen Maker: Mr. Chen":

> This is the story of one particular maker I have met in Shenzhen. It turns out, Mr. Chen was more interesting than his machine! You see, he has managed to carve out a nice little niche for himself by designing and manufacturing his own electronics and then selling them at the infamous Huaqiangbei electronics markets . . . as [we] got to talking about making and DIY culture, I began to get a sense that this down-to-earth guy was someone who really understands the so-called maker culture . . . what he was describing was a lean operation where he had digital fabrication tools that allowed him to retool and switch around really quickly and efficiently. His house was doubling as his production floor so he had very little over-head. He also understood that he needed to find niche markets in order to remain competitive.

This post framed Mr. Chen as an authentic maker, a man who had, out of passion, followed a lean approach to his electronics business. He practiced the kind of self-reliance, Smith explained, that "would turn any urban farm-loving hipster green with jealousy." Mr. Chen "lived" the kind of resourceful-ness that investors celebrate; following Mr. Chen to the rooftop of his house, Smith discovered that "Mr. Chen was not only running his own electronics manufacturing process, but was also growing organic vegetables, and rais-ing chickens and pigeons on the roof his apartment . . . To all the Mr. Chen's out there, and anyone else who pursues the goal of self-employment through making, I salute you!"

The idea that one should (desire to) "become self-reliant" was a mantra at the Shenzhen incubator, repeated over and over by the director, founders, managers, and mentors of the program. Stories of authenticity, of lean and fast production that (as I was often told) "the Chinese" simply know how to do, imbued a neoliberal rhetoric of self-reliance and self-investment not only with a sense of empowerment, excitement, and opportunity but also with a moral obligation. Shenzhen here is enrolled in the socialist pitch of entrepreneurial life; learning from Shenzhen, from Shenzhen speed, and its moral "model" makers was pitched as enabling the (re)appropriation of the "master's tools" of capitalist acceleration and speed.

When Audre Lorde famously provoked that the "master's tools will never dismantle the master's house,"[55] she suggested it was impossible to fight structures of domination if we depend on and benefit from them. For the people who the system continuously denies well-being and whose survival it divests—women of color, lesbians, and poor women, Lorde elucidates—radical opposition is not a privilege but the only possible survival. The "invention of Shenzhen" I document in this book was predicated on the promise that radical resistance was unnecessary, because the "master's tools" had become "democratized," made available to everyone. This promise of democratized intervention (via democratized innovation) was a depoliticizing move. It flattened across racial, class, and gender differences and implied not only that "everyone" was equally empowered by technological entrepreneurship, but also that "everyone" was disadvantaged by dominant structures the same way. Shenzhen, its economies of scale, its global supply chains and factories were enrolled in this "pitch" that resistant capacities were being democratized via open source technologies and commitments. Radical resistance, here, serves the function of the pitch—displaced into the innovator's toolkit of "disruption," it becomes a market device to lure feeling and serve the anticipatory logics of investment.

5

Seeing Like a Peer

HAPPINESS LABOR AND THE MICROWORLD OF INNOVATION

> Emotional labor . . . is the work, for which you are paid, which centrally involves trying to feel the right feeling for the job. This involves evoking and suppressing feelings. Some jobs require a lot of it, some a little bit of it . . . The point is that while you may also be doing physical labor and mental labor, you are crucially being hired and monitored for your capacity to manage and produce a feeling.
>
> —ARLIE HOCHSCHILD, "THE CONCEPT CREEP OF 'EMOTIONAL LABOR'"

"There is a thing about Sophie and Mei Yun.[1] They are the office managers. They make our lives better. They will make sure shit works around here. If it doesn't [work], just remember, it must be their fault. Repeat after me: 'nothing is Evan's or Mike's fault. It must be their fault. When shit is broken, talk to them!'" This is how Evan, the European director of the Shenzhen-based hardware incubator, kicked off the three-month-long training program in early 2013. It was the first week of the program, and Evan had ordered the members of the thirteen chosen start-ups to gather in the entry area of the small two-story apartment that housed the incubator. Evan's introduction to the two female Chinese staff was brief, but it set the tone for how the start-up teams would approach them over the months to follow: their job was, as Evan had made clear, to help guide the start-up teams through the emotional ups and downs of their entrepreneurial lives in Shenzhen. When "shit" was "broken," they would fix it. When the stress levels among the start-ups went up, Sophie and Mei Yun would provide distraction and plan fun leisure activities. The women's job was to stay in the background, to take care of the program's day-to-day functioning, and to ensure the start-ups' emotional well-being. From the first week of the program, then, Evan established that the role assigned the two women was one of "emotional labor," as it is defined by sociologist Arlie Hochschild[2]—the kind of paid job that involves the production of a particular feeling. Their job was to produce a feeling of "happiness" and cheerful delight—a feeling that was increasingly unavailable in the tech and design industries due to precarity, job insecurity, and broadening debates over the pitfalls of technological promise (see this book's introduction and chapter 3). These women were "happiness workers," providing a particular kind of affect and emotional support structure that made the work of the (mostly male) tech entrepreneurs not only bearable but pleasurable. Their labor made the precarious and high-risk conditions of "venture labor" both manageable and fun.[3] Sophie and Mei Yun, both in their early twenties and recently graduated from college, enabled and sustained the work of incubating human capital (see chapter 4).

In the previous chapter, I focused on the specific mechanisms that incubators and similar entrepreneurship training programs use to train people to self-invest; in this chapter, I concentrate on the gendered and racialized dimensions of this process. Specifically, I show how the promise of democratized innovation via peer production and open source enabled racial exclusion and gendered divisions of labor.

I found myself one out of only four women in the program (Sophie and Mei-Yun as support staff, me as ethnographer, and one female entrepreneur—my friend Serena). After the first few days in the program—a week that is engraved in my memory as one of the most stressful fieldwork experiences I have ever had, and one that I have long struggled to write about—I fell ill. My body shut down for three weeks, pained by continuous high fever and bodily inflammation due to a white blood cell increase, which my TCM (Traditional Chinese Medicine) doctor attributed to high stress levels. In that first week of the program, I had encountered an environment rife with the kind of masculinity now understood as pervasive in the tech industry and in computer science and engineering fields; I had also felt that I had to prove my worth to the program. As I explained in chapter 4, the incubator was an environment that induced desires for self-investment, which positioned participants as being in competition with one another.[4] It was training in how to "entrepreneurialize one's assets." As political theorist Wendy Brown articulates the ideology of neoliberal finance capitalism, it was an environment where "centralized authority, law, policing, rules, and quotas are replaced by networked, team-based, practice-oriented techniques emphasizing incentivization, guidelines and benchmarks."[5] Processes of self-economization[6] were couched in a rhetoric of coming together as peers and of hopeful change.[7] At incubators, people are trained to "pitch" a "better world" that is worth investing in. They learn how to see the world through promise. At this particular incubator, this meant framing the world through the promise of democratized peer production, community, and open sharing. This *seeing like a peer* reframed the world via hopeful stories of digital justice and of democratized technological intervention—of technology making the world a better place by making technological experimentation available to everyone.[8] Ironically, as this chapter shows, seeing like a peer also meant to unsee things that did not fit the anticipatory logics of investment—specifically, it meant to unsee gender and racial inequality and exploitation and to unsee potential for subversion and political action. Seeing like a peer framed everyone around us—including factory workers and staff—as peers; we were all in it together and we were all supposedly equal.

This chapter aims to contribute to the expanding body of research on digital/tech labor. This research documents the proliferation of labor exploitation and precarious conditions from factory settings to creative work.[9] Some of this research has been influenced by the autonomist Marxist conceptions of immaterial labor[10] and the social factory,[11] showing how capital subsumes various aspects of our lives for the accumulation of value.[12] By highlighting continuous and new forms of labor exploitation at the sites of digital production and consumption, scholarship on digital/tech labor has further dismantled techno-optimistic belief systems and raised important questions about the ethics and politics of digital platforms, social media, and Internet technology broadly. This work on labor exploitation in the digital age makes clear that the utopian visions of the global creative class,[13] of the network society,[14] and information society[15] were naïve at best and harmful at worst. It was the celebratory endorsements of Internet use and social media as inherently participatory, empowering, and democratizing, this work shows, that were deeply complicit with the creation of a lucrative online marketplace (that serves the likes of Amazon and Facebook),[16] with the advancement of surveillance capitalism,[17] platform capitalism,[18] and automating inequality.[19] I hope to further this body of work on digital labor by shifting some of the broad assumptions about labor in China and by revisiting our conceptions of subversion and intervention. This chapter shows that a priori associations with China as factory work or with labor exploitation in the factory occlude other forms of exploitation and precarity in China that have sustained the global tech industry. Specifically, I argue for the urgency to attend to labor that is necessary to nurture and support entrepreneurial living—of making the economization of life attractive and hopeful. I argue that attending to the gendered and racialized dimensions of this labor can challenge the notion that the economization of life happens due to some hidden process of the financial market. Human capital (and capital's subsumption of hope and promise) is not a by-product of neoliberal finance capitalism (nor of platform capitalism) but its main project; it has to be actively produced and worked on, it has to be incubated and nurtured. It is the feminized labor of maintaining the feeling of the good life that is necessary for the creation of human capital.[20] If we notice this labor, if we attach ourselves to the bodies who perform it (rather than to technological promise), we also notice that the economization of life is not an inevitable process—that it can be otherwise.

Happiness Labor and the Entrepreneurial Factory

During my ethnographic engagement at the incubator, I spent much time with two of the women in the program: Sophie, one of the Chinese female staff members, and Serena, the one female entrepreneur (Mei Yun, the other Chinese staff member, was out of the office for most of my time at the incubator). We spent time together inside and outside the office spaces, sharing personal conversations over lunch, dinner, shopping, and so on. The women rarely met as a group, but I spent significant time one-on-one with each woman, mostly during lunch breaks, over dinner, or during an occasional visit to the spa. The two women's different positionalities centrally shaped their ability to counter and push back on a variety of gendered and sexist commentary and practices at the incubator. Sophie, who had just recently graduated from college, was single, and came to Shenzhen to seek employment in the city's emerging digital tech, maker, and start-up industry. Serena is Asian American and had long-term work experience in the American tech and research industry; she holds a PhD and was in a committed relationship (her partner was the co-founder of her start-up and was also part of the incubator program). While Sophie had never traveled outside of China, Serena, like me, had traveled frequently for work and vacation to different parts of the world.

My own position also shapes this account. When I spent time at the incubator, I had just completed my PhD degree at UC Irvine and had begun a postdoctoral position at Fudan University in Shanghai. This was during a period when people both inside and outside the academy were critiquing higher education for its alleged distance and its lack of impact on the real world.[21] My peers had pervasive fears about future employment due to scarcity of academic jobs. The incubator promised a world where young people had a voice, where they could intervene in established structures from capitalist exploitation to the future of work; at this particular moment, that was incredibly enticing. But it was precisely this promise—the way that incubator programs, coworking spaces, and makerspaces (and academia) drew upon these yearnings for alternatives to existing structures of work and corporate control—that allowed old regimes of gender and racial exclusion to persist. In the incubator, a masculine, colonial discourse of adventure and control (that wove through its training materials and was enacted and "modeled" by the program director, the manager, and the at that time exclusively male mentors they invited) rendered China as needing to be tamed by (white) men. In other words, China itself was feminized and racialized; the investors and managers at the incubator had simultaneously leveraged and contributed to the "invention of Shenzhen" I describe in chapter 3—the oriental tropes of othering that were common among the designers, educators, and open source advocates who had turned to Shenzhen and that rendered China's "backwardness" as a unique opportunity for investment and for the recuperation of technological promise. This gendered and racialized attitude toward China was enacted in the regime of "happiness labor" required of young Chinese women (and at times of young Chinese men). This happiness labor sustained a myriad of incubator and start-up training programs in China that were modeled on the foreign-funded Shenzhen incubator (itself modeled on similar initiatives in Silicon Valley).

What I call *happiness labor* is the kind of labor relied on by a range of "new" organizational models of tech production—incubators, coworking spaces, makerspaces, hackerspaces, design labs, open innovation labs—to sustain the precarious conditions of the entrepreneurial life they demand and celebrate.[22] This labor provides the emotional and infrastructural conditions for flexible work and venture labor. It sustains entrepreneurial life by making its precarious and high-risk conditions manageable and "fun" for the creative and flexible worker. Happiness labor, both in China and in Silicon Valley, makes workers who are called upon to self-manage[23] and to self-invest[24] view these demands through an ethos of happiness. It allows them to sustain a "good feeling" about technology despite the increasing realization that for many, the self-investment will never pay off and that the promises of the tech industry have masked new forms of exploitation. Building on the sociologist Andrew Ross's work, the information and communication scholar Christian Fuchs argues that the precarious conditions of flexible work are "a form of absolute surplus-value production: one works long hours that under conditions of highly unionized and organized labor could look differently and these social costs are outsourced to individuals. The total wage and investment costs of those who make high profits are thereby minimized, which increases the rate of exploitation and profits."[25] People take risks and allow themselves to be exploited because of the promise that the Internet industry "has a certain, positive future," that people "had control and autonomy over their work and their choices in their industry."[26] Fuchs describes this as a "social form of coercion," stemming from "social pressure among colleagues, competition, positive identification with the job, a fun and play culture, performance-based promotion, incentives to spend a lot of time at the workplace."[27] He describes Google as the "prototypical implementation" of this ideological and social coercion "built into the company's culture of fun, play-bor (play labor), employee service, and peer pressure."

Despite this ideological coercion and pressures to perform, Google nevertheless provides for its small group of elite, highly trained engineers and programmers the working environment that can sustain this work. But this environment is mostly absent for most other workers in the creative economy and tech industry. The perks at tech companies like Google—cheap or free food, childcare, sports facilities, ball pits and slides—are longed for by many aspiring tech workers. And it is exactly the absence of such perks (the fun and play culture built into office space design and services offered at large tech corporations) that the young women (and some young men) employed as happiness workers help fill. Careful attention to this pervasive form of happiness labor makes clear the various forms of labor exploitation in the creative and tech industries. It is not only true that factory-like exploitation is seeping into various corners of our lives—what the autonomist Marxists called the social factory, but the inverse is the case as well; as I will show, factory work itself is increasingly shaped by the kinds of exploitation we see happening in the creative and tech industries—factory workers are called upon to see themselves as passionate entrepreneurial change makers, to turn themselves into assets on behalf of the company they work for and for the governments that invest in factory upgrades in order to render their respective regions attractive to future investment (this line of argument continues in chapter 6). What we witness, so to speak, is not only the expansion of the *social factory*, but the making of the *entrepreneurial factory*.

Happiness labor is precarious. It is low-wage labor, and often even unpaid, for its producers are promised that the experience will make them more competitive on the job market. Many of the young women I found employed in these jobs in China spoke fluent English, which they had acquired either by rigorous self-study or through college education. Many of them were severely underpaid. Despite the fact that many held college degrees, they were often paid about the same wage—at times an even lower wage—than the migrant workers employed in the electronics factories in Shenzhen. The exploitative nature of labor in the electronics industry has historically been rendered "worth it" and "camouflaged" by the techno-utopian promise of the digital artifact this labor produced—the smartphone, and particularly the Apple iPhone.[28] As digital race scholar Lisa Nakamura puts it,

> Apple effortlessly broadcasts the cachet of Silicon Valley design while camouflaging its industrial origins. It is made in China, but signifies to Chinese themselves, including those who are making it, that it is not Chinese, but rather culturally Other, and "cool." Apple has always highlighted its design origins in Cupertino (every Apple product comes with a piece of paper that states "designed in Cupertino, California") rather than its actual production culture in China in order to preserve the image of an artisanal product, a strategy which makes it especially and enduringly globally desirable.

A similar dynamic is at play in the happiness labor I observed. The exploitative nature of happiness labor was rendered invisible, its origins and production conditions occluded, in order to sustain the promise of the rise of a newly hopeful class of tech innovators who were celebrated as countercultural tinkerers who bring about social justice via the democratization of innovation. The young people employed as happiness workers wished to be viewed as contributing to this project of bringing about justice and social change by leading an entrepreneurial life.

As anthropologist Lisa Rofel shows, desire is a key mode in the experimentation and instability of neoliberalism. Throughout China's history of post-socialism, entrepreneurs have signified progress and migrant workers, backwardness.[29] As Rofel, building on feminist scholar Ching Kwan Lee, points out, factory workers in China display affective engagements not with the products of their labor nor with the job itself, but with future making: "the affective engagement here is with imagined futures." As Rofel argues, this is because "the dream of entrepreneurship" has extended into the labor of the factory: workers' "vision of themselves as entrepreneurs displaces previous notions of class identification as a proletariat and partially disrupts a view of themselves as part of a working class in the present."[30]

In this chapter, I show how a workforce of young, mostly female college graduates endured inequality in the present—not unlike the factory workers Rofel describes—as they dreamed of leading an entrepreneurial life, that would allow them—like the start-ups they worked for—to make a difference in the world. Attending to their labor challenges us to expand our conceptions of digital labor in China. While a rich body of work in anthropology, political science, sociology, and communication studies has documented with much nuance past and ongoing shifts in labor practices and exploitation in China, in the debates on digital labor China is often reduced to serving a particular argumentative function: China, and in particular the Taiwanese contract manufacturer Foxconn (which has its factories in China), is invoked, for instance, to show that nineteenth- and twentieth-century Fordist regimes of industrial labor exploitation continue in our age of post-industrial, postmodern, neoliberal, finance capitalism.[31] China is portrayed as the other, as a "less modern" site of surplus value accumulation that has persisted alongside the newer, more high-tech forms of labor exploitation embodied by tech companies such as Uber, Amazon, and Facebook and their "platform capitalism"—that while theorized as infiltrating China are seldom understood to be driven or co-produced by governance or the tech industry in China.[32] Stories of Foxconn workers' suicides are often taken to be emblematic of the conditions of labor in the Chinese tech industry. The obsession with Foxconn ironically re-affirms the Euro-American-centric notion that the non-West is in some way stuck in the past. China stands for old capitalism—pre-platform capitalism. China is limited to serving as a kind of atavistic survival of older regimes of industrial production and exploitation in the factory, while digital labor—immaterial labor, free labor, and platform capitalism—are theorized as originating from the West (or the global North). In the United States, taxi drivers commit suicide; in China, factory workers commit suicide. In Silicon Valley, young creatives and flexible workers struggle to pay their health insurance; in Shenzhen, migrant workers struggle to send remittances back home. The focus on a particular kind of factory work as paradigmatic labor exploitation in China occludes how the CCP has already been expanding the build-up of China's entrepreneurial factories for some time.

This is certainly not to say that careful studies of labor conditions in China's factories are not important. To the contrary, an excellent body of scholarly (and some journalistic) work in China provides crucial and in-depth insights for (but is seldom taken up in) debates on digital labor. This work shows, for instance, how contemporary labor conditions in China are deeply intertwined with shifts in global trade, geopolitics, and nationalist aspirations.[33] I follow this line of work to push for the importance of attending to the global and transnational circuits of the entrepreneurial factory that employs Asian women, not only as factory workers, but also as happiness workers, entrepreneurs, designers, producers, and administrative staff. When factory workers become a category, other forms of labor that are made to serve the entrepreneurial factory, and the dreams for agency and social change that sustain it, are rendered not only invisible but also seemingly non-exploitative.

This chapter focuses specifically on how dreams of social justice and yearnings for alternatives are not empowering for everyone in the same way. While some find empowerment, others (and especially those deemed less capable of becoming countercultural hackers) get caught in these acts of yearning—they get stuck in a seemingly endless and vicious cycle of dreaming up alternatives that make them endure the present.[34] I show that our scholarly frames of what counts as digital resistance and authentic Internet counterculture is complicit in the processes of the entrepreneurial factory. This chapter thus interweaves an account of the hopes for technological alternatives in technology research with accounts of the various dreams for justice I studied.

The Promise of Proximity

"At times I feel my job only consists of ordering things on Taobao for them!" Sophie said with exasperation. It was late in the day, and she and I had gone out for dinner to a small noodle shop around the corner from the accelerator office—I knew it was one of her favorites, as it served food from her home region. She insisted on inviting me along, and since it was an affordable RMB 20 (around USD 2–3) meal, I agreed. We both enjoyed our private get-togethers that provided time and space away from the pressures to perform in the office. We ordered our noodles and sat down at a narrow table at the window with a view to the passersby of the electronic markets' busy streets. "It's just so infuriating," Sophie said. "It's as if I don't even exist as a person." Earlier in the day, one of the start-up teams had sent Sophie a request to order electronics parts from the Chinese ecommerce website Taobao. Taobao was something the start-ups continuously marveled over for the speed of delivery ("what takes weeks [to ship] in the US, takes a day in Shenzhen") and the sheer scale of content sold ("you can buy anything on Taobao"). But Taobao was difficult to navigate for the mostly foreign start-up teams. Each transaction required the buyer to carefully check product quality and vendor trustworthiness, and the sheer mass of seemingly identical products with a broad range of price and quality demanded careful research, comparison, and sometimes lengthy chats in Chinese with vendors. Watching Sophie navigate the complex site and then inspect the quality of products when they arrived eventually taught the start-ups to purchase components on their own, but it was a tedious and somewhat boring form of labor. Discovering an unusual or particularly intriguing item in the Huaqiangbei electronics markets was exciting; in the earlier weeks of the program, these finds were demonstrated with much fanfare in the office and met with admiration from fellow teams and the program managers. But the work of sourcing components on Taobao was less flashy. The start-ups often rushed their orders or delivered their order requests to Sophie via email at the last minute. In this particular case, what had angered Sophie was that the start-up team had not even bothered to engage with her directly, but had emailed their order to the program manager, who then forwarded it to Sophie. "They wouldn't even talk to me or recognize I was there. This is not cool. I don't like teams like that," she told me.

We spent a good part of our dinner conversation talking about the office culture and about Sophie's future and her career options. Before she had gotten her job at the hardware incubator, she had worked for a foreign corporation that produced furniture in China. Sophie had found the work boring and isolating; her days consisted of fulfilling requests from her male boss, with little other social interaction. When the opportunity to work for the incubator came along, she was excited. When I asked her how she felt about it now, she evaded the question at first. She began explaining how she dreamed about running her own business, perhaps in trade, or becoming a member of a start-up team: "I really like working as a team." I asked if her experience at the incubator had positioned her well to start something of her own. "I am not sure if I could do it. But I really want to," she responded. "You know this is still better [than my old job]." She then added, "I might be able to go to Silicon Valley, to the United States, for the first time. But even if I can't go, this will be a win-win situation, because the start-up teams will be there [in Silicon Valley], and if I can't go, I get at least some time off." In the weeks leading up to the start-up teams' final product pitches in San Francisco (see chapter 4), the conversations between Sophie and me repeatedly returned to the topic of Sophie's potential travel to the United States. Sophie didn't know whether the program director "would allow it." He had been avoiding answering Sophie's inquiries about it, and her dream to travel to Silicon Valley was kept dangling, seemingly within her reach. She didn't even know if she could get the visa to travel. "You know how hard it is for most Chinese to travel abroad, right? Even for going to Hong Kong, I need a special permit. I can't just go like they do [the start-ups]," she told me one day as the two of us walked through the electronics markets to find a repair worker for a broken mobile phone. As we jumped across the potholes on the sidewalks, which were covered in dust from the months of construction work for the new subway station in the Huaqiangbei electronics markets, the busy-ness around us suddenly made me dizzy. The city's frantic remaking of its industrial sites into glossy images of creative "high-tech" work felt numbing. The seductive draw of future-making and the dreams of accelerated change felt like nothing but a façade that masked the hardship most people endured. Why do workers continue to endure in this way?

The incubator pitched self-investment as granting happiness—it would allow people to live the life of an "ethical" tech producer, committed to making the world a better place. When this happiness is dangled in front of people (like Sophie's trip to the United States), people endure more and for longer because they are promised to eventually become virtuous (and thus happy) change makers themselves. As feminist and critical race scholar Sara Ahmed reminds us, the promise of happiness works through a continuous postponement: "happiness is what makes waiting for something both endurable and desirable—the longer you wait, the more you are promised in return, the greater your expectation of a return."[35] Working for the incubator offered proximity to entrepreneurial life—happiness and the good life would be granted to those who had self-transformed into the kind of moral tinkerers who intervened in the status quo in order to improve people's lives and make the world better. This promise of proximity made happiness labor attractive to young Chinese college graduates like Sophie. The cruelty of this form of labor exploitation lies in the fact that the promised happiness seldom materializes.[36]

A couple of weeks into the program, a Swiss film crew set up their cameras at the incubator. The crew became a regular feature, documenting the day-to-day workings of the incubator and following the start-ups: on factory tours, to the markets, when they tested their products in the wild, and eventually, on their trip to Silicon Valley. The film director was particularly interested in one of the Chinese start-ups, a group of young men from a second-tier city in the western part of China. The documentary that was eventually produced, called *The Chinese Recipe*, shows with much care the struggles of the young Chinese men striving to gain legitimacy in a Western-centric tech industry where racialized stereotypes leveled against them were common—they aspired to be taken seriously and understood as just as innovative as their Western counterparts.

Like Sophie, they dreamed about going to Silicon Valley and meeting some of the widely celebrated American maker advocates whose work had motivated their own in China. Only some of the team members received Chinese visas to travel to the United States, but those who did fulfilled their dream of traveling to California and meeting their maker idol, Chris Anderson. The film ends with a shot of the young men at a beach in California, looking outward over the open water toward the horizon. This was not a celebratory story of success, the film carefully suggests. We are left with a sense of ambiguity, as the film neither celebrates nor critiques the young men's seeming fulfillment of their Silicon Valley dream.

I first watched the movie in a hotel room during a research trip to China in 2016. I had met with the film director, whom I had come to know well, over dinner, and he had brought me a CD of the finished film as a gift. As I watched the film, there I was, included in the story. The ethnographer was caught in the act, notebook and recorder in hand, suspiciously eyeing the charismatic aura assigned to the white man as I followed the Chinese team, who stood in awe when Chris Anderson granted permission (with an attitude of annoyed benevolence) for their encounter with him, so longed for, to be filmed. Who is granted proximity to objects of happiness, to the promise of future making? Who is kept at bay?

Sophie did not go to Silicon Valley with the start-ups that year, a topic that she tended to avoid toward the end of the program. "Someday, I will go. Someday, for sure," she told me the week before the teams, the film crew, and I prepared for our overseas trip.

The Microworld of Innovation

"I really wish there were other women in this," Sophie told me often. At the time of my research, the mentors chosen for the accelerator were exclusively male. While there were women in the manufacturing industry, most of them were assistants—positions that were sometimes powerful, responsible for managing international relations and national trade networks, but that were nearly always backstage. Neither Serena, the one female entrepreneur admitted to the program that year, nor Sophie could turn to a female role model in this space.[37]

The day I landed at the airport in Shenzhen in early January 2013 to begin my research with the incubator, I received a text message from Serena: "Lady, get your ass over here quickly! This is such a sausage fest!" When I arrived a couple of hours later at the office, Serena's words rang in my ears. Upon entering, I was hit by a strong smell that reminded me of my early forays into the worlds of engineering and the tech industry in Germany and Austria—male deodorant and aftershave mixed in with the smell of burned plastic, machines, and paper. Loud voices came from around the corner. I walked further in to say hello. A group of (mostly white) men stood laughing around a table that held a large piece of equipment. One of them looked up and glanced at me, then returned to the conversation. The rest had taken notice too but chose not to interact. Uncertain what to do, I decided to act as if I was cool with the whole situation. I leaned on one of the tables close to the window, looking out onto the spread of Shenzhen's buzzing electronics markets ten floors underneath. I pulled out my phone and texted Serena, asking where I could find her, my heart pounding. As I debated whether I should gather up my courage and walk over to the group of men, the doorbell rang, and Sophie came down the stairs from the second floor. She looked at me, smiled, and said before she opened the door, "Oh, hi, are you Silvia?"

Only later did it dawn on me that I had entered a twisted version of what historian Paul Edwards calls a "microworld." Building on STS scholar Sherry Turkle's work with engineers at MIT,[38] Edwards describes how

> members of many computer subcultures—hackers, networking enthusiasts, video game addicts—become mesmerized by what Turkle calls the computer's "holding power," i.e. the computer's ability to fascinate, to command a user's attention for long periods, to involve him or her personally. For many there is a deep form of pleasure that derives from a blending of self and the machine, creating a microworld, i.e. a unique ontological and epistemological structure that is simpler and seemingly more manageable than the world they represent.[39]

Edwards explains that these microworlds are appealing because they are "free of unwanted complexity." Within them, "things make sense in a way human intersubjectivity cannot," and "the programmer is omnipotent."[40]

The incubator functioned as such a microworld, extending masculine ideals of control and mastery to China, its people, and its markets, factories, and supply chains. The incubator's foreign founders, investors, and mentors and its director and manager were crucial to this process of displacing old ideals onto "new territories." This was a microworld in which the enactment of a particular form of masculinity—one that extended technological mastery beyond the "safe" world of the geek and hacker and into the concerns of finance investment—was central to demonstrating competency.[41]

Early in the program, the management team had organized a conference aimed at introducing the new batch of start-up teams to Shenzhen's transnational start-up and maker community. That conference set the tone for the months to follow. About sixty or so people gathered in the entry hall on one of the upper floors of a slick modern high-rise in the western part of the city, just a thirty-minute car ride from some of Shenzhen's largest manufacturing towns. There was excitement in the air, a sense of being part of something greater. The building in which we had gathered still reeked of new paint and sawdust. It was as if the city was getting ready for these thirteen start-ups to put their touch on what China's tech industry aspired to become.

We all sat down in a room with rows of chairs in front of a podium where the speakers had already set up their presentation slides. Evan, the program director, took the podium to kick off the event. The organizers and speakers wore casual clothes—shirts with eclectic prints, jeans, sneakers, a wearable electronic gadget here and there—the cool clothes that promise the socially awkward nerd confidence. Serena and I stood out against this background, simply by not wearing jeans and a t-shirt. On his head, Evan wore a pair of electronically powered pink and white bunny ears with sensors attached to his forehead. The sensors signaled each ear to move forward and sideways based on brain activity that was invisible to us. With his bunny ears wiggling at seemingly random times, giving off a tiny mechanic sound you'd imagine a cute animal robot to make, Evan, the investor, looked like a geek. The invited speakers he had gathered were "amazing folks," he began. "Amazing folks, who I have pulled from all over the world to come here and talk," people who had all been in various ways "engaged with the ecosystem here in Shenzhen." Before turning to the first speaker, he added: "Hopefully I'm proving that you can just be normal and not socially awkward wearing bunny ears. And it's great, right? Because someday we will all wear them. Awesome!"

The incubator, Evan's speech made clear, would be an exciting adventure driven by people who wanted to help make radical transformations in technology and society. "I want to chat about what's happening, here [in Shenzhen], really, today. It's pretty amazing," he said. "If you go back in time, we are currently re-watching what's happened in the 1960s/70s. Of course, hardware has been here for a while now . . . but this time is different . . . what you see really is an uptick in a lot of variables going in the way of the entrepreneurs." He characterized the accelerator as a mentorship and training program that would enable the daring to transform themselves in order to reach new worlds of innovation. According to him, the start-ups in the room were not only following in the footsteps of the earlier generation of idealized countercultural hackers but also taking on the task of disrupting old and "outdated" industries, moving them to the next level. The promise was that *everyone* would have the power to become an innovator like Steve Jobs. The difference was that they would hack not only technology but also—and perhaps more important— markets, supply chains, and the tech industry as a whole. As Evan declared, the program was "really an experiment," both for the foreign VC firm Evan represented and for the start-ups themselves. The VC firm had been investing in software start-ups for several years. Their move into hardware was risky, but potentially highly lucrative. Prototyping the future of hardware entrepreneurship in China would, they hoped, justify the investment.

Evan continued: "you have people here in China that bring it down to business and make it happen. So, this is about what's really happening in Shenzhen today. So, we are really boldly going where no man has gone before." This casual reference to the American pop-sci-fi TV series *Star Trek*'s famous opening lines ("Space: the final frontier. These are the voyages of the starship *Enterprise*. Its five-year mission: to explore strange new worlds. To seek out new life and new civilizations. To boldly go where no man has gone before!") aligned maker and geek culture—and the interest of capital investment—with a masculinist colonial narrative of adventure, exploration, and technological mastery. As "a narrative of American frontierism, exploration, and 'boldly going'" critical computing scholar Paul Dourish and anthropologist Genevieve Bell argue, "*Star Trek* portrays space as the final frontier and technology as a tool placed in the hand of individuals who are accorded the freedom to wield it."[42] As a "collective imaginary,"[43] *Star Trek* has shaped how people in the tech industry and research think about future technologies and who they consider central protagonists. Implicit and explicit references to *Star Trek* reframe the complex issues that arise at the nexus of science and society in terms of a techno-centric discourse that makes them appear solvable through technological solutions.

While the uptake of the *Star Trek* narrative is not unique to this particular hardware incubator program (it is a common narrative in technology research and design more broadly[44]), it served a particular function. Let me unpack briefly. The investors, the program director, and the manager of the incubator program made frequent references to *Star Trek*. For example, when one of the employees of the VC firm visited the program to give a lengthy lecture on how to build a successful start-up, he argued that the start-up was a team on a mission that required "the best people on board." He exhorted them to consider how "taking the starship *Enterprise* on its mission was never a one-man job, and the captain alone would have been helpless. OK, so he had a vision to 'go boldly where no man has been before . . .' but he needed Mr. Spock (analysis) to make calculated choices, Scotty (systems) the engineer to maintain the machinery, and Bones the doctor (people) to keep the team in good health. And so it is in start-up land." This reference to a team with different specializations positioned the start-up as capable of managing the complex social and material processes of industrial production in China by working together. As the program manager repeatedly emphasized, "hardware is hard"—much harder than software. The start-ups were trained to channel their commitments to democratize innovation into the labor of pitching, managing relationships to Chinese factories, vendors, and supply chains, and following international trade and product safety regulations. These *Star Trek* references rendered China, and its complex social and material infrastructures of industrial production, as "controllable" and "manageable" if the geek could channel his commitments, transform himself, and develop his ability to wield both technology and markets and economies at scale. If space was the final frontier to be mastered by individuals empowered by the latest technological advances, China was the untamed land of hardware innovation to be mastered by the start-up. This required not only a technological vision, but the proper execution: an ability to turn makers' dreams of hacking alternatives into economic realities.

A question often posed to the start-ups by visiting journalists (and a question I was often asked when I presented my research) was whether they were worried about their ideas being copied and/or stolen by "the Chinese." Most start-ups replied that this was not a concern, because despite the ability of "the Chinese" to outperform their Western counterparts in speed and scale of technology production, they did not—so the argument went—know the specifics of what made Silicon Valley tick. As the program managers made clear, the start-ups' competitive advantage was in expertise and know-how in building platforms, smart hardware for IoT, and data-driven companies— things attractive to investors. This kind of "advanced" know-how was viewed as being (still) absent in China. Colonial tropes of China as somehow backward endured in this translation of a (masculine) mastery over machines into a mastery over a region and its supply chains, factories, and "a Chinese culture of copycat." The endurance of such masculinized forms of othering is not unlike what feminist STS scholar Christina Dunbar-Hester observes in American radio counterculture; "people's dissent from 'mainstream' values," she shows, "is no guarantee of a rejection of masculine enactments of technology."[45] On the contrary, stipulates STS scholar Ron Eglash in a similar vein, "geek culture serves a gatekeeping function, shutting out non-whites and women."[46]

Why do these old masculinist and colonial narratives of othering and exclusion gain new currency in the very moment that many in the tech industry sharply critiqued them? We can begin to understand the endurance of old regimes of power and control when we turn to the dreams and hopes for alternative futures that make us unsee inequalities in the present.

We must pay particular attention to the historical specificity of how and when these dreams emerged. The incubator program capitalized on young people's dreams of building concrete alternatives to exploitative tech and corporate control in the tech industry. Tech work was becoming feminized; the skill content of the work was being devalued, and pay was therefore falling—a process historically associated with the introduction of female labor. Maker discourse of the period was saturated with a nostalgia for the past of valued (and financially valuable) craftsmanship and "authentic" (rather than automated) tech production. Colonial and masculinist tropes endure in these promises of empowerment that frame certain individuals as capable of regaining control amidst such processes of feminization. In the narratives of the incubator, people would increase their market value and regain a sense of control if they could learn from, and then learn to master, Shenzhen's "craft-like" (feminized) ingenuity and informal production culture. Shenzhen was portrayed as an opportunity to re-live a past where men mastered, owned, and were in control of the technologies they made. Such forms of empowerment have always been less available to women and people of color. Women, for instance, were historically excluded from craft work coded as masculine—craftsmanship rather than handicrafts, which have always been portrayed as the feminine domain.[47] As craftsmen formed unions to fight capitalist management and deskilling, they actively excluded women in order to defend their work, feminist STS scholar Judy Wajcman explains:

> Craft workers were an elite group who enjoyed a privileged position in the labor market and considerable autonomy over the labor process. [. . .] Labor process and Marxist theory focused on relations between worker and capitalist, which underestimated divisions within the working class such as those based on sex, race, age, and skill in shaping the effects of technical change on the workplace. Feminist writers have exposed the inappropriateness of the craft model by highlighting the exclusivity of craft unions as male preserves . . . Craft workers who have been seen as the defenders of the working-class interests in struggles over technical change in part derive their strength from past exclusionary practices. Their gains have often been made at the expense of less skilled or less well-organized sections of the workforce, and this has in many cases involved the exclusion of women.[48]

Like craft labor, technology is not inherently masculine; as feminist scholars have shown, the associations of masculinity with technology and machinery are socially constructed,[49] and that social construction is historically situated. As Wacjman elucidates, craftsmen who were fighting to protect and secure their conditions of employment created and enforced skill designations, then defended those skills by excluding others. In other words, men's resistance to capital has long operated against women's interests. Defending skill—preventing "dilution"—has almost always meant blocking women's access to an occupation.[50] Contemporary processes of feminization in the tech industry have motivated new fears that the often male-dominated industries and fields in computing will be decimated by automation and artificial intelligence; these anxieties sustain masculinist enactments of technology, turning them from labor-based to technology-based enactments, and they sustain the colonial and gendered discourses of technology innovation. It was in this moment of renewed fears over loss of control, which were particularly pronounced in the American tech industries (see introduction), that many turned to Shenzhen, portrayed as the opportunity to relive the masculine and colonial dream of control. Ironically, the happiness labor of young Chinese women like Sophie was crucial to this dream.

What Makes an Authentic Hacker

Sophie was frequently asked to organize and accompany factory tours, which were crucial for the start-ups. Not only did they help to build relationships with potential partners and clients in manufacturing, they were central in producing the kind of affect of interventionist capacity that many in the thriving maker scene had come to associate with Shenzhen; engagement with Shenzhen promised not only to open up the black box of end-consumer products, but also to open up the black box of economies of scale, acceleration, and speed. Many of the maker advocates I met over the years explained that their main motivation in traveling to Shenzhen was to "see" the scale of industrial production "in action." Shenzhen factories were considered an opportunity to re-engage with the kind of production expertise that the West had given up or "lost" due to outsourcing and automation. Factory tours were thus crucial in establishing an affective relationship between start-ups and factories—it was work, I learned, that was performed by a whole flurry of young Chinese men and women, employed not only by this particular hardware incubator, but by both the Chinese state and corporations, often free of charge. This work, a labor that supported affective ties, also helped legitimize Westerners' engagement with Chinese factories, formerly decried (often by the same people who now wished to learn from them) as sites of labor exploitation and industrial excess. The tours offered a glimpse into the scale of Shenzhen's industrial production and the various decision-making processes occurring on the factory floor. The tours were thus a crucial node in crafting Shenzhen's allure as allowing a going back in time to prototype alternatives to the exploitative tech industry (see chapter 3). The start-ups' engagements with factories gave them an aura of authentic hacker expertise (they were hacking capitalism "right there on the factory floor"), the return of which was so desired in American and European maker networks. (This was a marketable skill on its own; after completing the incubator program, many graduates worked as consultants or in jobs that invested in the promise to open up the inner workings of industrial production.)

The production of this authentic hacker image required that an emotional bond be established between the start-up and the factory, between the foreign entrepreneur and China. The labor that went into building this affective bond (performed by happiness workers like Sophie) required the ability to relate both to the mostly male factory managers and their interests and to the motivations and ideals of the mostly male makers and start-ups. This labor is not unlike what communication studies scholar Nancy Baym has termed "relational labor" in the music industry.[51] Social media platforms, she shows, have brought with them new demands on musicians' relations to their fans. Drawing from Arlie Hochschild's seminal work on "emotional labor,"[52] Baym demonstrates that social media, originally used by creative professionals to expand their audiences and market, now demands relational labor: "the ongoing, interactive, affective, material, and cognitive work of communication with people over time to create structures that can support continued work." In other words, the musician must perform not only creative labor but also the emotional labor of managing fan relationships, which they must treat as deeply personal, free of money.

Sophie and many other young college graduates like her performed a type of relational labor in their work with start-ups: they created affective communicative structures that supported the making of an authentic hacker subjectivity (and its economization by venture capital), but (unlike musicians), they did so *on behalf of others*. The musician might eventually benefit economically from their relational labor, but happiness workers are a mere proxy; they are reduced to the function of platform, of infrastructure, for their job is to allow others to relate to one another with the kind of affect necessary to get the job done. In a perverse twist, the affective tentacles of entrepreneurial life and platform capitalism reach into the factory floor—co-producing the entrepreneurial factory.

One of the many factory tours Sophie organized was a visit to a facility with CNC drilling and laser cutting machinery. This tour was similar to many other factory visits, but also differed because the large-scale machinery this facility offered, especially its laser cutters, were especially popular sites among the start-up teams. As one of the male start-up team members excitedly told me before the visit, the large CNC machines were the "real deal," unlike the smaller low-end versions that equipped most maker- and hackerspaces. Mike, the manager of the incubator, had decided that this tour would be relevant to all the start-up teams. The day before the proposed tour, he had tasked Sophie with organizing it—a job that required her to allocate travel time and match the start-ups' schedules with the factory schedules, all of which required advance planning. "Tomorrow?!" Sophie had replied. Several hours of phone calls later, she walked past my desk, rolling her eyes. "Ugh, it's done." She then walked from desk to desk, reminding all the start-up teams to gather outside the office building the next morning at 8:30 am. Responses ranged from "That early?" to "Hell yeah, lasers!" The subway ride would be long; it would take us about 2 hours to reach Longhua, at the end of the red line. The next day, a small group (not everyone had made it out of bed in time) got onto the subway, quiet and red-eyed. Sophie had purchased a bag of steaming buns at the corner of the subway stop, and she began to pass them out. "Here," she said, handing the bag to one of the start-up teams next to her. "Omg baozi!"

The group was suddenly in better spirits. Two hours later, we got off the subway and hit our first stop: a small coffee shop near the station that Sophie had looked up. "So needed this coffee," said Serena next to me. As we slurped our hot drinks, Sophie was on the phone, coordinating with the driver who would pick us up to make the drive the additional twenty minutes from the subway stop to the factory. A black Mercedes SUV and a silver Volkswagen waited for us outside the subway station, and we squeezed in—the people outnumbered the available seats in the two cars. "Ladies on top!" Mike screamed, grinning at Serena and me. Serena grimaced at him. The guys giggled.

We squeezed into the car, Serena squashed into a corner with her partner, Sophie and me awkwardly teetering half between and half on top of the guys. The driver of our car, who turned out to be the factory manager, didn't speak English. Sophie started up a conversation with him, referring to him with the common and somewhat respectful phrase "laoban"—boss. She inquired politely yet casually about how business had been lately, how long the factory had been operating, about their customers and trade partners—gleaning knowledge that was crucial for the start-ups to know, which she would convey to them on the subway ride home. The laoban was talkative and wanted to know what brought these "laowei" (the foreigners) to China. Sophie explained that we were ambitious young "chuangke" (makers) who were working on "really innovative things." "And they are really excited about seeing the machines," she added with a big smile. The laoban laughed. "Chuangke, eh?" He smiled. "Good, good." One of the team members picked up on the Chinese term for maker and chimed in approvingly, "Yeah! Chuangke!" Everyone laughed. The banter continued, with Sophie translating for both sides and adding explanation for each if needed, and the twenty minutes in the car went by quickly.

The car stopped just beneath a big street sign: "西门富士康." Sophie translated for the teams: "Westgate of Foxconn." "Ah?!" somebody responded. Sophie responded patiently: "We are in Longhua. That's where Foxconn is located." We had turned from a big two-lane road onto a narrow, unpaved side street. A two-story building complex ran along the street. Through its large garage door–style openings, I could make out bodies and machines at work. We had arrived at our destination: a smallish facility filled with various types of CNC machines. Here, the men at work were moving between software interfaces and the machines. "Wow, this is awesome!" somebody yelled as the group walked further in. "They have only seen smaller machines," Sophie explained to the manager with a chuckle. She added, "Some of them might need your service for their IoT products. And they love machines like that." She continued her small talk with the factory manager, then turned back to the start-ups and said in English, "So what questions do you have for them? You could ask about their clients, their workflow, the materials they use and so on."

FIGURE 5.1. A CNC drilling machine—a machine of the size and capacity that sparked admiration and fascination among the start-up teams. Photograph by the author.

What ensued was an exchange between the start-ups and the factory manager that was carefully calibrated and negotiated by Sophie, who switched between English and Chinese, adding in a joke here and there for each side. The visit lasted more than two hours and concluded in the factory owner's tearoom. Drinking tea with the owner or manager was a common and important part of doing business with factories in Shenzhen. Sometimes the type of tea offered initiated conversations about the factory owner's background and home region; most of the factory owners and managers I met in Shenzhen had come to the city from somewhere else for a better job and life. Most were male. The conversations were not always easy, as regional dialects and drastically different backgrounds sometimes made it difficult to sustain the kind of prolonged informal conversation that is so crucial for establishing trust and bonding in the industry. Having been on visits like these without Sophie, I knew exactly how tiring and exhausting it was to do that kind of relational labor between start-up and factory boss. It required not only translating between English and Chinese (handling many unusual Chinese words and phrases), but also keeping up with the kind of cheerful talk and happiness labor demanded by interactions with the start-ups.

Seeing Like a Peer

Over time, the kind of happiness labor that was part of Sophie's work became normalized among the start-up teams. They began relating to her and talking about her as a friend and a peer, framing her labor as the kinds of favors that you'd ask of a good friend. In other words, they saw her labor as a form of peer production. "Sophie didn't feel like a staff or a translator to us. She was like a friend, a peer," one of the start-up members told me when I asked him years later about Sophie's role in the program. "She was like a friend, somebody who can tell you about the culture of Shenzhen, for instance how the locals do certain things, dealing with the factory. She helped us navigate the daily life in Shenzhen and to get our stuff actually done." However, the start-ups' revision of Sophie's role from a worker to a peer masked the gendered division of labor at the incubator, and the uneven positionality between the firm's employees and the entrepreneurs it invested in. This reframing of Sophie as a peer and a friend empowered some of the men to demand that Sophie extend her happiness labor and perform other favors for them. For example, some of the men asked Sophie to bring along some of her "Chinese women friends" to the after-work gatherings she organized, and more explicit requests that Sophie introduce the men to "Chinese girls." Eventually, Sophie organized some gatherings and outings that included some of her friends, but she was angry about it: "Not sure why I am doing this. They are not serious about dating any of us." She was right; a couple of days later, I stayed late in the office, after Sophie and most others had gone. I overheard two of the guys joking about how "these women" weren't "marriageable material." This was about "having some fun."

Sexist jokes surfaced repeatedly in the office, accompanied by loud laughter and boasting. I struggled to extract myself from these conversations. One of the start-up teams was working on a smart sex toy for women, and Evan, the program director, took this as an opportunity to ask Serena and me our "expert opinion" on the subject matter. I remember it vividly. It was a quiet day in the office, with several of the start-up teams tapping away at their computers, and Evan suddenly stood in front of me, loudly demanding an answer: "So, given we have some ladies here in the room, I was thinking we could get some expert opinion here. Silvia, how would you describe your sexual preferences?" He paused there, grinning, looking around the room. His statement had attracted attention. He continued, "This is really about what women want, not about some company dictating their desires." Sexual harassment as women's empowerment?! Stunned, I sat silent, searching for the right words. Serena jumped in before I could think of a reply. I was struck by how at ease she seemed in the situation. In a strong voice, sounding confident and in control, she first mocked Evan and then responded sincerely to the question, treating it like a technical matter. I remember thinking to myself that her response portrayed her as an equal to the men (and as different from the other women) in the room. In the microworld of entrepreneurial life, some women could at times enact masculinity in order to gain a voice, respect, and an aura of authenticity.

Years later I was again reminded of this incident. At the 2016 Shenzhen Maker Faire, I had paused at one of the event stages, where a fashion show was under way. As I watched the twenty or so slender women in glamorous dresses, enhanced with blinking LEDs, walking up and down the stage, I noticed that right behind me, people were gathering. The crowd was surging around a woman with her hair tied into ponytails encircled by LEDs, her large breasts embellished by a chain of LCD shatter glass, a skirt barely covering her thin body, and black platform boots. She appeared to be the fleshly manifestation of the hypersexualized machine-woman perfected for the male gaze. Her name was Naomi Wu (@RealSexyCyborg). As she posed, imitating the models a few meters away on stage, a crowd of mostly men surrounded her, gazing, grinning, pointing their cameras. She was like an exaggerated mirror, a parody of the events unfolding on the runway: women would be allowed on stage, but not to speak, only to be gazed at. As she would explain this event later, "In . . . 2016 there was a Maker Faire held here in my home city of Shenzhen. Not a single Chinese female Maker was invited. In order to protest this, and Maker Media's ongoing exclusion problem in the most visible possible way, I built a 'Blinkini' . . . While the wearable is risqué, a more low-profile project would not have made an effective protest since it would simply have been ignored." I was intrigued—I could not help but wonder whether Wu had masterfully pulled off the feminist intervention of parodying and mimicking normalized structures of injustice, making them visible through exaggeration;[53] by mirroring and distorting what was happening on stage, she unmasked the continuous sexism of the tech industry that positioned women on the sidelines, their bodies objectified, their voices excluded, rendered less technical, less valuable.

Naomi Wu's Twitter account, @RealSexyCyborg, has—at the time of writing—115.9 thousand followers. What I had witnessed at the Shenzhen Maker Faire in 2016 was the first of a series of public performances that Wu carefully crafted for immediate effect and documented (often by herself) via video for later online consumption. These acts were largely aimed at an English-reading audience, broadcast on a YouTube channel and on Twitter (both of which are blocked in China, accessible only via a VPN). Naomi Wu had created a public persona in @RealSexyCyborg, known for calling out Maker Media and the tech industry writ large for their continuous exclusions along gender and racial lines. When Dale Dougherty, founder of Maker Media, tweeted in 2017

FIGURE 5.2. 2016 Shenzhen Maker Faire Fashion Show (left) and next to the stage, Naomi Wu surrounded by a group of men taking photographs of her (right). Photograph by the author.

that Naomi Wu herself, like @RealSexyCyborg, was "a persona, and not a real person," he provided fuel for her critique. Her Twitter followers revolted and Dougherty officially apologized and featured her on a cover of *Make:* magazine several months later. Many have written about this incident, including *Vice* magazine, which Wu subsequently accused of journalistic malpractice. *Vice* responded to the allegations, stating that while "Wu's activism" was "hostile and aggressive," there was an "organic authenticity that both Wu and *shanzhai* undeniably emit." When Dougherty said that Wu was "not a real person," he provided the ideal ground in which to grow Wu's (and many others') demands for changes to the status quo. With astonishment, I watched many people taking Wu's side: people who had never been to Shenzhen, people who did not know much about the maker movement, people who had never heard about Naomi Wu prior to the incident, including colleagues in academia—all of them denounced Dougherty for his racist and sexist response. Several of the men at the Shenzhen hardware incubator, including both mentors and participants, also spoke up on her behalf. What made them see in Naomi Wu's case precisely the things they chose not to see at the incubator?

There is—as *Vice* claims—something authentic, something familiar about Wu. She acted like a powerful man; she demanded attention, demanded control of the conversations about her, demanded to be heard. She was aggressive, assertive, and hostile. She self-sexualized. She parodied dominant norms of resistance—demanding, forceful, masculine. She was authentic. *Vice* magazine's comparison between Wu and shanzhai helps elucidate my point here. As

I show in chapter 3, shanzhai has undergone a remake: what once represented China's supposed inability to innovate is now seen as a "clever survival tactic" that confronts and subverts Western hegemony. Silicon Valley appears to feed capitalism; shanzhai appears to hack it. Shanzhai, the Chinese copy, recuperates counterculture; copy is re-scripted as a more authentic hack. Wu's performance was seen as authentic, because it appropriated the West's pervasive yearnings to resist the tech industry's old regimes of power and their broken promises. @RealSexyCyborg was noticed and celebrated, exactly because the hacktivist could recognize "himself" in her (despite or perhaps exactly because of her sexualized, hyper-erotic body). What Wu—deliberately or not—exposed is how, despite mounting criticism of Silicon Valley—from the proliferation of labor exploitation through tech platforms such as Uber and Amazon to the industry's pervasive sexism—entrenched notions of what counts as resistance and who counts as authentic hackers remain. @RealSexyCyborg was seen as subversive because she appropriated an already legitimized form of resistance embodied in the male countercultural hero and hacktivist. Her performance felt familiar to her followers on Twitter and YouTube, exactly because she reflected the West's own tendencies to imitate and constantly repeat its own idealizations of true, authentic innovation, resistance, and counterculture.

While Wu was celebrated, the exploitation that Chinese women employed as happiness workers experienced continued to go unnoticed. But does this mean that the Chinese women employed as happiness workers merely enabled the machinery of finance capitalism? Is there subversion in happiness labor? And how would we notice it?

On to another factory tour. In this facility, the working conditions were particularly bad. Unlike the large contract manufacturers like Foxconn that are now emblematic of harsh factory labor conditions in China, these small- to mid-size facilities (which have received little censure from the Western media) often lack some of the most basic health and safety measures for their workers—measures that the bigger contract manufacturers cannot afford to skimp on. As we entered the building, a harsh toxic smell made my eyes tear up. As we walked further in, I noticed that the young male workers who were coloring and sanding plastic encasings were wearing no protection over their eyes or mouths. In another room, a group of women had just returned with food to their workstations, where they began to eat amidst machinery, soldering stations, and fumes. Throughout the entire visit, not one of the start-up teams commented about the conditions. Instead, they all gathered in the machine room, marveling over the size and speed of the equipment. Some of us stepped out of the building to wait in the courtyard for the rest of the group. Sophie walked away and squatted on the unpaved, dusty ground to wait. "What the hell are you doing?" one of the American guys screamed. "How can you even squat like that?" Sophie gave no response and continued to squat there

FIGURE 5.3. October 25, 2016 Naomi Wu posted on Imgur a photo of her intervention at the Shenzhen Maker Faire 2016, with the following description: "This project was inspired by my experience with *Make:* magazine. While I'm a huge fan they have a pretty serious dress code for women appearing in the print edition—they have not shown a female midriff in eight years. This has made it difficult for the current generation of young female Makers who work on Wearables, Cosplay and Fashion Tech to have their work acknowledged. Make has refused to discuss just what the editorial guidelines are, so we have no way of adjusting our content to meet them . . . Given what I'd be going through with *Make:* I knew I wanted something involving electronically variable modesty, as a kind of defiant social statement (boys are disruptive with their tech skills all the time, it stands to reason girls would be as well—with the issues that are important to us) . . ." (excerpt). From Naomi Wu, http://www.thingiverse.com /thing:1848578, CC BY 3.0

patiently. After what seemed like a brief moment of deliberation, he squatted next to her, struggling at first, then finding his balance. "This is hard!" he yelled. Laughing, two of the other guys followed suit. Sophie remained silent, vaguely smiling. There seemed to be nothing to add; the silence had a certain power over all of us, squatting quietly on the dirt road.

"This was a hard visit," somebody said as we walked back to the subway stop ten minutes later. Serena turned to me and asked, "Are you gonna write about the working conditions in this factory?" It was the first time I had seen the start-ups openly acknowledge the labor conditions in the factories. Many of them—like most maker advocates I met in Shenzhen over the years—had talked about factory labor only in the language of peer production and shared expertise. They often told me about instances where workers had helped them solve complex problems that had arisen during the production process, about processes that workers had developed to speed up production, and about the workers' tacit knowledge of materials, which they thought they themselves lacked. As one prominent hacker once told me, "The problem with the factories is that they don't compensate workers for this expertise." He told me about an experience from a visit to a factory that produced clothing. One of the factory's line workers had developed a special mechanism for holding the piece of fabric, which not only sped up the production process but also improved the quality of stitching. The worker, however, had not benefited—in terms of his pay—from this invention.

Over the years, I listened to many such stories and read many blog posts and travel reports from maker advocates who had made their way to Shenzhen that were depicting—often in detailed photo essays—factory workers as their peers and close collaborators. In these accounts, the worker is celebrated as the locus of shared expertise, a peer and co-producer of tech innovation, an ally in fighting the establishment. And yet, unlike the maker advocates who benefited from being seen as treating factory workers like partners, the workers depicted remained unnamed and their work conditions unchanged. Seeing like a peer and unseeing inequality thus displaces possibilities for collective solidarity. The feeling that they "hacked" capitalism on the factory floor in Shenzhen made the start-ups and maker advocates who celebrated workers as peers unsee their own complicity in labor exploitation. "I didn't *see* labor exploitation during my time in Shenzhen," one of the start-up members insisted during a follow-up interview I conducted with him years later. Others explained to me—ironically—that challenging labor exploitation in a Chinese factory (which required political rather than technical competencies) was beyond their reach and abilities.

Sophie's act of quiet squatting had created a temporary crack in this otherwise seemingly hopeful, depoliticized attitude of *seeing like a peer*. Anthropologist Anna Tsing argues that it is in the gaps of the otherwise all-encompassing and seemingly smooth operations of capitalism that alternatives emerge.[54] To account for these cracks, she argues, we need to practice an art of "noticing differently." Feminist scholarship suggests that we seek out these temporary glitches, for these imperfect, compromised, ambivalent gaps allow us to notice alternatives that are worth pursuing. Sophie's squatting was odd and ambiguous, impossible to celebrate as a countercultural act of resistance, and yet it constituted a rupture of the smooth surface of her job, which demanded that she make people feel at ease. It resembles the everyday acts of resistance—the "weapons of the weak"—that anthropologist James C. Scott documents. Squatting, Scott argues, like other resistive acts such as "foot-dragging, false compliance, feigned ignorance, desertion, pilfering, smuggling, poaching, arson, slander, sabotage," is part of the "small arsenal of relatively powerless groups," an "ordinary act of class struggle" employed in moments "when open defiance is impossible or entails moral danger."[55]

The squat was a subversive act, I argue, but one that escapes the typical definitions of such. Rather than an attack upon the system (as if a weapon), it was a more ambiguous and partially complicit act; while it seemed to instill some reflection among the start-up teams, it is neither clear what role Sophie's particular actions played in triggering their questions, nor if it was a deliberate act on Sophie's part. It also did not call into question the incubator itself or its workings (at least not directly). But it is this ambiguity, the act's seeming inconsequentialness that deserves attention, for it can tell us something about the structures of power in which it is embedded. In the contemporary political economy, feminist scholar Anna Watkins Fisher shows, powerful entities—from the "'open' corporate platform" to the "'welcoming' security state"—control not by direct coercion but by promising open-ness, participation, and hospitality. In this economy, "disruption and critique are not what threaten the stability of the system but are essential to its functioning" and "critics find themselves implicated in the very power structures they seek to oppose."[56] Let's return to Naomi Wu's performance to help elucidate this argument; her parody that mounted to a demanding and forceful protest against sexualized and racialized discrimination of the tech industry constituted a productive force; it "pitched" @RealSexyCyborg as an authentic (female) hacker who stood up against the system and won (by bringing her likes and acclaim, especially on Western social media); and it served Maker Media itself which, by including her on the cover of *Make:* magazine could reclaim its commitments to open-ness and democratized innovation. Wu's branding of herself as an authentic hacker ironically fueled the socialist pitch that runs the entrepreneurial factory; the promise that individuals have the capacity to change things occludes how systemically nothing much has changed at all; the exploitation of Chinese women in factories, in incubators, in coworking spaces, in the sex industry continues on, in part because some Chinese women (and some Chinese men) are now granted the status of radical hacktivist. What we are left with is the affect of change—a feeling that somebody fought the system and won, while the system itself remains intact.

This is not to argue, Fisher urges, that resistance is altogether impossible, but to rethink it. "Is resistance thinkable," she asks, "from a position that is not autonomous but embedded?"[57] Fisher suggests yes, but this resistance is parasitical—ambivalent, non-emancipatory, and compromised. That resistance is embedded does not imply total subsumption, she argues further; instead, parasitical resistance is "slow resistance" that "signals possibility" by "incrementally redirecting the host's resources from inside."[58] Both Naomi's and Sophie's actions could be read as versions of Fisher's parasitical resistance, each in their own ways chipping away slowly, ambivalently, at the system that sustains them. But while Naomi's performance is publicly recognized and celebrated as countercultural resistance, Sophie's micro-performative act was short-lived and is easily brushed off as odd—it's challenging to make sense of it. It is in its ambiguity, in its seeming futility, its noncompliance with "pitching"—with being celebrated as "radical" and "countercultural"—wherein we can notice a gap in the seemingly all-subsuming spiral of capitalism. Historian Michelle Murphy argues that "as capitalism captures desires for a better world . . . one might be tempted to see capital as driving all wants."[59] To work against this totalizing narrative, we must attend to the acts that fit neither the story of countercultural rebellion nor the story of endless subsumption. It is in these fleeting, ambiguous, and perhaps—just like Sophie's—odd acts (a glimmer rather than a prototypic alternative[60]) that dents and gaps become apparent in the functional alignment of dreaming of better futures and the anticipatory logics of finance capitalism.

6

China's Entrepreneurial Factory

THE VIOLENCE OF HAPPINESS

Robin was the first shanzhai entrepreneur I met.[1] He was—like me at the time—thirty years old. He had moved to Shenzhen from a rural part of China in 2003 at the urging of a relative, who helped him obtain a fellowship at the Chinese car manufacturer BYD (Build Your Dream). More than just a symbolic name, the BYD fellowship had enabled Robin to follow his dream of leaving behind what seemed like an already set future in agriculture—he found footing in Shenzhen's thriving electronics industry. By the time I met him, Robin had made a name for himself by producing a device that came to be known as the first shanzhai Apple iPad, released on the Chinese market before the first-generation Apple iPad hit the stores in the United States (for a definition of shanzhai, see chapter 3). When I first interviewed Robin, he insisted that this product had never been a copy. Rather, it was Apple that had copied him, he explained, elaborating in detail the timing of his own production and first release date. I stayed in touch with Robin over the years. In 2016, I visited his office at the outskirts of the city, where many small- to mid-size production houses were located. He had invited me to meet over tea to tell me about an exciting new business opportunity.

A childhood friend from his hometown had opened a shoe factory in Ethiopia. The idea was that Robin would get a section of the factory to have some of his own products assembled there. The shoe factory's promotional materials that Robin showed me signaled more than an attractive business venture. Echoing some of the CCP's own official rhetoric, the brochure spoke of mutual growth, a win-win situation, and postulated China's rapid economic development as a model for Africa. However, there was no direct government

FIGURE 6.1. Drinking tea at Robin's office at the outskirts of Shenzhen in 2016. Office (left) and building exterior (right). Photographs by the author.

support for the venture in Ethiopia, Robin explained. The expectation was that China's entrepreneurs—people like Robin and his friend from Hunan—would "bootstrap," would invest their own (human and economic) capital to "go out" or "go abroad" (出去 chuqu). While the personal and professional risks could be high, government discourse portrayed "going out" as carrying promise to redeem China and its place in the world—from a region known as fake and low-quality to cutting-edge and high-tech innovation. This was possible, so the promotional materials of numerous government-sponsored events on chuqu I attended that year made clear, if Chinese people upgraded themselves—just as Robin was hoping to do: from a "low-quality" farmer over a "low-quality" shanzhai producer into a modern, cosmopolitan, globally savvy business man. Robin described it to me as "a matter of feeling responsible (责任感 zerengan)." This opportunity, he explained, was in fact a burden (and as I will show, a mandate): "In Shenzhen, people push you to go further . . . but our responsibility, our mission, also becomes a burden. We are responsible for many workers, and this makes us feel we cannot stop, cannot let go."

What Robin describes here, a feeling of responsibility to improve the self on behalf of others, is also the central focus of this chapter. Specifically, my aim is to show how the government's appropriation of making was aimed at portraying the demand placed on individuals to refashion themselves from "economic" actors into "entrepreneurial" actors as serving the interests of the people. Making, I show, was convenient for the government's aims to build China's entrepreneurial factory, i.e., the making of a machinery that would instill desires for human capital—individuals who understood self-upgrade and self-investment as providing them with the agency to prototype themselves and the nation anew. China's entrepreneurial factory runs on the promise of individual redemption tied to national redemption; if people like Robin—people who had made themselves into successful economic actors that contributed to China's success of rapid economic development—managed to upgrade themselves into the kind of "entrepreneurial" dreamers that carried international cachet and stature, they would also redeem China on a global stage as a forward-looking and happy nation, seen as promising by others. This chapter shows the two processes that were key to the build-up of China's *entrepreneurial factory*: first, the stipulation of certain people (both Chinese and Westerners) as prototypes of this cosmopolitan worldly citizen who modeled this self-transformation, *and* second, the deliberate upgrade of select urban spaces and infrastructures in order to create a sociomaterial milieu that modeled happiness and the good life, thus inducing desires in people who did not (yet) have access to these spaces to self-upgrade—a process I call *contagious happiness.*

When I returned to Shenzhen in 2018, two years after Robin had first told me about his planned ventures in "going out," he had moved to a new office, located in one of Shenzhen's newest and most prime real estates—a newly built "technology ecosystem" park in Shenzhen Bay (深圳湾)—the same district that had hosted Shenzhen's second featured Maker Faire in 2015, which was also the first Maker Faire in China that received sponsorship, official branding, and endorsement from the government. As we sat in his new office, thirty floors up in the modern high rise overlooking a pristine golf course landscape, Robin showed me the video conferencing system he had set up and the photos from his latest trip to Ethiopia and Kenya. The new office space stood in stark contrast to the office Robin had occupied until just a year ago; its high-tech conference system, the modern light wood furniture, the generous room sizes, and the floor-to-ceiling windows gave it the kind of creative feel typically associated with "new" spaces of innovation: coworking spaces, IT office spaces, incubators, and their open office floor designs and affective "playbor" aesthetics.[2] As I complimented Robin on his success and the new space, he interrupted me, his voice getting quieter than usual. "This is not it!" he stressed. When I asked what he meant, he explained, "I envy you! You are respected, no matter where you go. When I go to Africa, or if I went to the United States, I am not respected. That's why we [the Chinese people] still have aspirations, that's why we hope to be respected. This is my goal. I am pursuing this goal, that's why—although today I could have a really good life—I am still living the life of a loser (屌丝)."

What weaves throughout this chapter are accounts of the slow violence in the form of continuous postponement of the good life, the endurance of racism and colonial othering, and the normalization of precarious and

FIGURE 6.2. Visit to Robin's new office in Shenzhen Bay in 2018. Office (left) and building exterior (right). Photographs by the author.

high-risk entrepreneurial life that the very project of the entrepreneurial factory demands and yet masks behind the promise of happiness.[3] The promise of happiness, feminist and critical race scholar Sara Ahmed reminds us, is an instrument of neoliberal governance that charges individuals with the "responsibility to be happy for others."[4] Happiness, she argues, is a "redescription of life as a project."[5] It has a promising nature, suggesting that happiness lies ahead of us, at least if we do the right thing, attach ourselves to the "right" objects, the "right" dreams, and the "right" people.[6] Making was ideal for the purposes of the CCP to posit certain people and certain spaces as the "right prototypes" of China's indigenous innovation economy, of the Chinese Dream, a happy nation. It was via the appropriation of making—and its associations with happiness and agency—that the CCP could "pitch" itself as acting in the interest of the people, as committed to people's pursuits of happiness. In other words, the CCP appears as a benevolent, optimistic "peer" of the people who works right in step with people's dreams and hopes for justice. Under Xi Jinping, the pursuit of happiness, feminist scholar Jie Yang shows, has been rendered a moral imperative for the "quality" citizen. The CCP—more commonly associated with top-down authoritarian control—governs (select groups of its citizenry) via the promise of happiness. Happiness might be granted, for instance, if (and only if) people like Robin find ways (on their own) to "go out" to places like Ethiopia on behalf of the nation to work "alongside" the CCP and its ambitions of infrastructural upgrade and reposition China globally via the Belt and Road

Initiative (BRI) and the "made in China 2025" initiative. Building up China's entrepreneurial factory, i.e., inducing in people feelings of responsibility to upgrade themselves into happy, entrepreneurial change makers on behalf of the nation, would provide the labor (and the human capital) of "going out" that was necessary to guarantee the success of these nationalistic infrastructure projects aimed at repositioning the communist party state on both a national and global scale.

In this chapter, I draw from interviews and observations I conducted between 2013 and 2018 with the workers, designers, engineers, and low-level managers in China's electronics production and trade industry. Some of these were people who came to Shenzhen in the 1990s and 2000s, in lieu of or after high school or college, and who built their careers in Shenzhen's informal grey market electronics production—through shanzhai. Others were younger—Chinese college graduates in their mid to late twenties and early thirties, who either came of age in Shenzhen or moved there to work in the tech/design/industrial production sector. I had conversations with people about their own life stories, about Shenzhen's histories of electronics production, and about China's stature in international relations. The people I met don't form a homogenous group and cannot be clustered by generational differences, class differences, transnational experiences, and so on. My aim here is to retell their stories—the struggles and aspirations of the people who allowed me to follow them.

Making Space for the Entrepreneurial Factory

In the spring of 2015, the Shenzhen city government erected a ten-meter-long, three-meter-high concrete sign at the bottom of the SEG market, one of the most iconic buildings of the Huaqiangbei (华强北) electronics markets. The sign, in both Chinese and English, read "深圳 — 华强北 — 创客的摇篮创业的天堂" or "Shenzhen Huaqiang North Commercial Circle—Cradle for Entrepreneurs Paradise for Starting a Business." It was put in place only weeks before Shenzhen hosted its aforementioned second featured Maker Faire. It was no coincidence that the sign was erected in the heart of the Huaqiangbei markets, which had been at the center of the "invention of Shenzhen" that I chronicled in chapter 3. It was one of the first arrival destinations for newcomers who wanted a firsthand feel for what many in open hardware and maker circles had begun referring to as Shenzhen's unique "innovation ecosystem"—a place to get in touch with the scale and speed of mass production, a place to "feel" and "see" capitalism "in action." Huaqiangbei spans several city blocks of the Futian district, comprising roughly a 15-by-15-city-block area. The area is made up of department store buildings, some 20–30 stories high, filled with tight labyrinths of stalls. Vendors sell everything from new and recycled small electronic

FIGURE 6.3. Signage in the Huaqiangbei electronics markets, spring 2015. Photograph by the author.

components to finished products—mobile phones, tablets, hoverboards, selfie sticks, security cameras, and much more.

The sign that the government had put up in the heart of the markets captured with just one phrase the government's intents and purposes when it had endorsed making just a couple of months earlier: it celebrated "makers" who were "entrepreneurial," and it celebrated Huaqiangbei as the incubating site of that renaissance: "a cradle for makers, a paradise for starting a business." That same year, I noticed that certain sections of the Huaqiangbei electronic markets were being redesigned. I didn't even notice the changes at first, since the markets had looked like one big construction site since 2013, when a new subway stop was being wedged into the already busy public transport grid beneath the electronic markets. The soundscape, day and night, had been the harsh noise of heavy cranes lifting materials through the air and of large machinery digging into the street. Walking on the streets of Huaqiangbei and past the street-facing vendors on the first floors of the buildings was treacherous, as construction materials blocked the way and dusty roads made sudden potholes disappear in a foggy haze. The new subway was highly controversial, for many of the owners of the market buildings feared losing merchants and business.[7] But during the subway construction, the interior of the electronics markets had remained largely the same. In 2015, this began to change. I first noticed these changes when I returned to Mingtong (明通), a mall known for offering the most curious designs of shanzhai devices. On the mall's third floor, through an intricate labyrinth of tiny stalls showcasing the latest offerings of electronics, packed with a smorgasbord of cardboard boxes, electronics, tea kettles, computer screens, fans, and magazines, there had been hidden a narrow escalator ascending to the mall's upper floors, which had held some of the most extravagantly designed shanzhai phones, produced in small quantities and with prices ranging from USD 15 to USD 40—the devices that had made the scale and speed of shanzhai itself felt (see chapter 3).

In 2012, I had begun collecting some of these devices. I purchased phones in the shape of Hello Kitty figurines, car keys, Chinese alcohol bottles, and cigarette boxes; pink and golden phones the size of a 5mm thick pocket calculator to fit the tiniest purse, phones the size of a brick with golden casings, phones that came with a solar panel on their back for mobile charging. There were phones whose shapes served a special purpose, such as the phone that also functioned as a flashlight and radio or the phone for the elderly with an enlarged keyboard. The highlight of my collection was the phone I ended up calling "the Buddha phone." It came in a wooden box whose insides were lined with a golden silk fabric. The phone is round and ornate, with a green jade stone decorating its back and a Buddha statue engraved on its front. Each accessory, from the charger to the software installation CD to the headphones and stylus, appears carefully designed to match the design of the phone. The software interface, too, reflects this style, featuring a modest and subdued green-yellowish graphic design and various chimes for meditation. The phone was Buddhism packaged as an "out of the box experience"—the kind of designerly approach to open source hardware that Massimo Banzi had described as differentiating Arduino from Chinese copycats—or "Chinaduinos" (see chapter 3). As I learned that year, the Buddha phone and the numerous other shanzhai phones were considered by many, including their manufacturers, to be relics of China's past, a stepping-stone in China's path toward becoming a leading global high-tech production house.

By the spring of 2015, the Mingtong vendors of these devices and the tiny escalators that had led to them had vanished. Puzzled, I wondered if I had misremembered the building or the exact location or both. It took me a couple of runs up and down the mall to realize that the two hidden floors were being renovated, being replaced with an open floor design. Gone were the mall's most hidden corners and the quirky, open designs that have become associated with Shenzhen's shanzhai economy.[8] While the mall was still filled with strikingly similar items, the look and feel had drastically changed into what seemed more like a mash-up between a modern design exhibit and an Apple store than an international hub of electronics trade. The rows of small stalls had been replaced with larger storefronts, rooms separated by glass walls, each with shelves displaying a neatly arrayed line of products that glimmered in the indirect light of rows of LEDs. The market's sociality had shifted too. Gone were the young couples with children running between the stalls and the small kitchen nooks that had served them. The showrooms were now

FIGURE 6.4. A security guard in SEG Market inspects a small-size 3D printer. Photograph by the author.

serviced by teams of young sales staff with professional attire and attitudes. The hallways were wider; the lighting and colors gave the space a look and feel of "legitimate" innovation rather than an impression of copycat culture. Gone were the hidden floors and crowded alleys of stalls that had hinted at a thriving grey market economy.

It was not only Mingtong mall that was being upgraded. SEG, one of Huaqiangbei's most iconic markets, had gone through a similar transformation. One of its lower levels now featured a new showroom that displayed its products with a designerly DIY appeal, lending the products an aura of playful high-tech innovation. Products were neatly arranged on white shelves and in museum exhibit–like showcases. In contrast to the previous market interiors, which had jammed a flurry of products into the small, cramped displays of tiny vendor stalls, here each product was given air and space, put on display, often labeled with the designer's or design firm's name. These products' branded identities hinted at intellectual property; they included globes that hovered above a surface, product lines of 3D printers and drones, household appliances, smart watches and virtual reality glasses, and robots that circulated through the aisles, approaching curious visitors with an endearing, child-like manner. The new showroom also incorporated a coworking space that served coffee and offered tables at which to linger and chat. Decorated with cardboard boxes, this part of the showroom conveyed a vibe of DIY making and unhinged creativity.

That same year, the municipal government sponsored the arrangement of a makerspace on the thirteenth floor of the tower of the aforementioned SEG market. SEG makerspace officially opened in the evening of the first day of the 2015 Shenzhen Maker Faire and of the city's first Maker Week, both of which were funded by organizations strongly tied to central, municipal, and provincial governments. The 2015 maker events marked a departure from the earlier, fairly small-scale maker-related events in Shenzhen, drawing a large number of international open source hardware and maker advocates in addition to visitors from municipal government offices, manufacturing, design, and tech. The hardware incubator discussed in chapters 4 and 5 was now located in the building next door to SEG, owned by the State-Owned Enterprise (SOE) Huaqiang. The new space was about ten times the size of the incubator's old offices and positioned the incubator in the heart of the electronics markets. Huaqiang also set up its own new space, called Huaqiang makerspace, on the top floor of one of its key buildings. Its main entrance was located on the building's rooftop, in a large outdoor patio space with lush greens and murals depicting sci-fi–like imagery of machine-centric future living. Entering from the patio, one arrived in an open, light-filled space with a coffee bar and high tables on one side and a product showcase on the other. From there, down a hallway, one glimpsed studio spaces with large glass windows and doors that allowed visitors to get a quick view of their inner workings.

None of these making-induced urban upgrades, however, had taken over the markets completely. Instead, only dedicated spaces and floors were redesigned. This "dual system" was aimed at inducing what many described to me as a "maker spirit" on a broader scale—desires to upgrade the spaces and people that had not yet refashioned themselves into entrepreneurial makers. This "dual system" of "old" structures, principles, and norms *side by side* with a set of "new" norms and practices will sound familiar to scholars of China; as I introduce it at greater length in the first chapter of this book, the government had strategically deployed a similar technique during the 1980s economic opening reforms, when it deliberately opened up to capitalist market processes only a select number of designated special economic zones (Shenzhen itself being one of them) that co-existed alongside the "old" socialist system. This "dual system" created competition over access to resources and—what political scientist Mary Gallagher calls—a *contagious capitalism* by inducing desires for social and economic upgrade and decreasing resistance to the "new." Just as the dual system of the 1980s produced desires in individuals to refashion themselves from socialist into economic subjects, so does this contemporary dual system (via select maker-induced urban upgrades) induce desires in people to self-upgrade from economic into entrepreneurial subjects—individuals are promised to gain the status of forward-looking, innovative, happy, and world-class Chinese citizens. While Deng Xiaoping stipulated in the 1980s that it was ok "for some people to get rich first," the current regime implies that it is ok "for some people to get happy first." While the CCP had long dangled the redemption of China's people as modern, quality citizens like a carrot (see chapter 2), this time happiness appeared within reach—there was a sense that Chinese people had agency to decide for themselves what their futures would look like.

What brought about this sense of agency was the socialist pitch of making itself—the promise (as articulated by prominent American maker and open source advocates) that becoming a maker meant acquiring the ability to intervene in the status quo via technological experimentation; political change and justice for the people was brought about not by collective mobilization or political action, but by the democratization of individualized tinkering. Some Chinese people had seemingly already accomplished this; as I have documented in great detail in chapters 1–3, China's early maker advocates were celebrated by the international maker scene *and* by the CCP as innovators that China could be proud of—the kind that built reputable international businesses (rather than copies) and the kind that adhered to international values of innovation and creativity (be they the principles of open source or the principles of intellectual property). It was this labeling of some people as prototypic entrepreneurial makers (by the CCP and the international scene of making alike)—during public events at Maker Faires or in prominent pieces of writings circulated through Chinese and American media outlets—that induced desires in many others to refashion themselves in similar ways. The maker-induced urban upgrades paired with the making of prototype citizens created an entrepreneurial factory that ran on contagious happiness.

A Lucrative Business—Investing in China's Innovators

In June 2015, two maker-related events took place in two different districts of Shenzhen, each sponsored by the government entity that represented the respective district. In the district of Shekou, the *Shenzhen Maker Faire* received government support from the SOE (state-owned enterprise) China Merchants, which has managed the urban and economic development of Shekou since 1979. In the district of Futian, China's first *Maker Week* was held in the Civic Center and was co-sponsored by the Shenzhen municipal government. The municipal government had insisted on the June date for both events in order to coincide with the US Maker Week being hosted at the White House by President Barack Obama. Many of the foreign maker and open source hardware advocates invited to present their work were confused by the decision to split the events between the districts of Shekou and Futian, which were separated by an hour subway ride. Many complained to me that the physical distance prevented them from seeing friends or particular shows and talks at the other event in town. They supposed that the strange location of the events was due to the lack of advance planning, which was considered common in Chinese society. But their assumption about this supposed lack of "Chinese organization and planning" precluded them from seeing just how much deliberate planning had gone into the orchestration, timing, and location of the two events. Each of the two government entities organizing and sponsoring the events was working to promote its district (and that district's industries) as uniquely positioned to implement the government's 2015 "mass makerspace" directive. And each district (and the SOEs associated with it) framed its own idiosyncratic urban and industrial development over the past forty years as what made them ideal for the advancement of the central government's interests. But why would city districts compete rather than work together on implementing the government's maker directive?

When the Shenzhen SEZ was established in 1979, the central government divided it into smaller development zones, each managed by a different SOE that was in charge of land development and industrial development in their zone.[9] In the 1980s and early 1990s, these SOEs operated with great independence from central authorities, each experimenting with different urban forms and modes of living, often drawing inspiration from drastically different urban development models in Hong Kong, Singapore, Taiwan, and other regions.[10] Urban studies scholar Ting Chen describes this as an economic and urban strategy of "self-evolvement," meaning that Shenzhen was built in the absence of, rather than through, central authority. Shekou, where the Maker Faire was held in 2015, constitutes one of these smaller development zones (an area of 10.5 sq km), held and managed by the SOE China Merchants.[11] SOEs like China Merchants, each of which owns and manages a designated plot of land in the city, compete for central government resources. A particular zone or district (and their SOE or political leadership) that demonstrates the successful implementation of a central government directive (like the "mass makerspace" initiative) can expect either the advancement of the local political leaders' careers or increases in regional investments, and sometimes both.

For our purposes here it is important to convey the wealth and power that SOEs like China Merchants have accumulated over the years by transforming themselves into real estate businesses. In the early 2000s, when the earlier mode of urban expansion that had characterized the SEZ in the 1980s and 1990s had reached a limit, a shift toward upgrade and renewal occurred within the SEZ. These "upgrades" focused often on transforming "outdated" urban structures and neighborhoods[12] and were largely aimed at maintaining Shenzhen's "'special' political position and [its] political privileges in finance, taxation, and land administration," amidst rising investment into cities like

Shanghai.[13] The municipal government created a series of incentives for SOEs to take on this kind of real estate development. SOEs thus transitioned from "pioneer builders" that tested different sociospatial approaches during China's opening in the 1980s into wealthy real estate companies that "expanded aggressively over time into new territories in Shenzhen or in other parts of China."[14]

For SOEs like China Merchants, then, demonstrating that their respective districts or zones were at the forefront of implementing the central government's maker initiative was lucrative; supporting district-specific maker initiatives allowed SOEs to brand their most recent wave of urban renewal projects as cutting edge, innovative, and in line with the interests of the central government, which in turn attracts investment to their districts (from the central government, corporations, and investors)—all of which increases the value of the SOE's real estate. Making was ideal for assigning value to real estate development projects as it required only minimal financial investment—especially in contrast to the costs associated with building up new industries or infrastructures—and provided "content" attractive to residents and local business alike.[15]

Whereas in Shekou, the SOE China Merchants lead the district's ongoing upgrades, in Shangbu, where the electronics markets of Huaqiangbei are located, several smaller-scale SOEs (e.g., Huaqiang, CEC, and AVIC) are managing the respective land. Urban upgrade in Shangbu thus unfolds through continuous negotiations between these various SOEs. In contrast, because of Shekou's comparatively large tract of land and the fact that it is managed by only one SOE, the look and feel of the urban space is drastically different from the urban fabric of a place like Huaqiangbei. At the seaside, with lush greenery that surrounds a myriad of luxury low-rise apartment complexes interspersed with high-rise buildings, Shekou represents China's aspirations for high-tech futures and modern living. Huaqiangbei, on the other hand, on a much smaller parcel of land managed by multiple SOEs, long had a very different urban aesthetic that was conducive to the build-up of the vast network of small shanzhai entrepreneurs in the late 1990s and 2000s, but is considered by the municipal government as representing the city's past. The maker-induced urban upgrades of Huaqiangbei I described earlier have to be understood as centrally shaped by the SOE's and their districts' competition over resources and investment.

When the SOE China Merchants sponsored the 2015 Maker Faire venue—Shekou's new "industrial software park" (in a neighborhood referred to as Shenzhen Bay 深圳湾)—it intended to signal that the central government's take on making was to be found not in the markets and tiny streets of Huagiangbei but among the modern high-rises, lush greenery, generously spaced software parks, and expensive housing complexes of Shekou. Looming over the venue of the Faire was the still-under-construction headquarters of China's Internet giant Tencent, designed by the American architecture firm NBBJ (which

FIGURE 6.5. 2015 Shenzhen Maker Faire in Shenzhen Bay, with the Tencent building, still under construction, looming over the Faire. Photograph by the author.

has designed other tech firms' headquarters, including Google, Samsung, and Microsoft). It is a vertical campus that stretches into the skies of Shenzhen rather than across the ground—an interconnected architectural design designed to reduce social hierarchies.[16] The building appears to enshrine into material form the aspiration of cultivating technology producers who would aim for the skies. When I arrived at the Maker Faire venue in Shekou, I was astonished. Screens several stories high, mounted onto the buildings of the software park, were announcing speakers and visitors from both China and abroad. Colorful Maker Faire banners floated in the wind beneath skybridges connecting some of the taller buildings. The sheer street space covered with maker symbols and artifacts made this Maker Faire one of the most grandiose I had ever seen. In contrast to some of the prominent Maker Faires I had attended in the United States (SF, Detroit, New York, DC), which aimed at giving off a vibe of DIY and self-made creativity, the government sponsors of this Maker Faire aspired to something else: the aim was to brand Shekou as capable of implementing China's entrepreneurial factory.

Two of the buildings that had been completed by the time of the Maker Faire were slowly filling with the kinds of innovative spaces the government had hoped would proliferate: incubator spaces and coworking spaces alongside individual office space occupied by tech start-ups and design studios. The Chinese open source hardware company Seeed Studio (discussed in chapters 2 and 3) had moved parts of its sales and design division into one of the new buildings. The other half of the floor adjacent to Seeed Studio was a large space filled with cubicles, empty except for one start-up occupying a single corner. In the building next door, a similar floor layout and empty desk spaces promised future productivity. By sponsoring the Maker Faire, China Merchants was breathing life into the newly built innovation park that housed the Faire. Like other events that are strategically positioned in parts of the city slated for urban redesign (for example, the Shenzhen Bi-City Biennale of Urbanism\Architecture (UABB)), the 2015 Maker Faire was in part meant to demonstrate that Shekou (and with it China Merchants) was the ideal district and organization to implement the central government's maker directive and thus receive further investment.

The—to the 2015 Maker Faire concurrent—Maker Week in the Futian district was not short on opulence either. The event opened with a spectacular ceremony at the city's convention center, located adjacent to the seat of the mayor and the municipal government. Chinese media were covering the event, and a select number of international and Chinese guests had been invited to witness the occasion, including Neil Gershenfeld from MIT (Massachusetts Institute of Technology), Clay Shirky from NYU (New York University), Lyn Jeffery from the IFTF (Institute for the Future), Tomas Diez from the Barcelona fab lab, Duleesha Kulasooriya from the Deloitte Center for the Edge,

FIGURE 6.6. David Li, and Neil Gershenfeld sign memorandum of agreement for Fab lab Shenzhen. Photograph by the author.

Tom Igoe from NYU and Arduino, Eric Pan from Seeed Studio, and David Li, co-founder of China's first hackerspace (see chapter 2), who had just taken the position of director of the Shenzhen Open Innovation Lab (SZOIL). The ceremony was brief; its highlight was the onstage moment when Neil Gershenfeld and David Li signed a memorandum of understanding. The ceremony sanctioned Shenzhen as a "fab city" and a "city of makers"—phrases that were blazoned across large banners and digital screens promoting the event alongside the city's roads and freeways and on corporate buildings.

In his opening talk at the following day-long conference, Gershenfeld spoke about how the current ways of doing things in technology and industrial production were passing into obsolescence. He described this as "a historical moment" in which what we had understood as modern technology (which he dated back to the work of Claude Shannon, John von Neumann, and Norbert Wiener) was no longer modern. He explained that this current moment was unique, for machines were on their way to becoming as flexible and modular as software. "It's like inventing an Internet for the visible world. That's the historical moment. It's really important that the goal of fab labs is to make themselves obsolete . . . In this world, where cities like Barcelona make what they consume, you don't need Shenzhen anymore." At the core of this transformation was what Gershenfeld's fab lab initiative had been promoting for some time now: "machines will make machines, fab labs will make fab labs, and cities will make cities." Shenzhen was crucial in this process of making existing infrastructures—from educational institutions like MIT to the global mass manufacturing embodied in Shenzhen—obsolete. Indeed, in Gershenfeld's vision, Shenzhen would make itself obsolete, just as Gershenfeld was making his own profession of university teaching obsolete. Shenzhen was providing the building blocks for a future of globalized automation machinery—of self-making and self-upgrading machines. The end state, he projected, would be a world where everyone would become "exactly the same"—just like "the inventive people at MIT":

> In Shenzhen, there are all these building blocks for machines that make machines, so there is this really interesting pivot from mass manufacturing to mass manufacturing of the building blocks for local manufacturing as a transitional economic statement. In turn, that stage then means Shenzhen is helping make the machines that make the machines. But in turn, it is empowering globally, but also in China. In manufacturing today, you will leave home to work in a factory doing something you don't want to do, designing something by somebody you don't know for somebody you will never see, to get money to then go buy something. If you can just make the thing, you don't need to do all of that stuff . . . Ultimately, if anybody can make anything anywhere and be connected, what's emerging in rural African villages and above the Arctic Circle is that we are finding exactly the same bright, inventive people you have at the MIT, but now on a scale of the planet.

The audience appeared ardent, and during and after the Q&A, several people specifically thanked Gershenfeld for his inspirational talk. The people who I spoke with afterward agreed that Gershenfeld's vision was promising for both Shenzhen specifically and China writ large. Why was this proposal to make existing structures obsolete so enticing? Why did the municipal government ally with Gershenfeld rather than a Chinese university partner?

Gershenfeld was invited by Shenzhen's municipal government for several reasons. He articulated a connection between Shenzhen and what the central government itself was promoting at that time—industrial upgrade (via initiatives like Made in China 2025). Gershenfeld pitched industrial upgrade not only as inevitable but as desirable; everybody who participated (including China's people) would become "as inventive as the people at the MIT." In China, the MIT has long been held in high regard, and many Chinese educational institutions have worked to compete with the MIT's perceived technological excellence and innovative exuberance. A partnership with the MIT would thus have provided cachet for the Shenzhen government no matter who was standing on stage. But Gershenfeld's particular vision that promoted self-upgrading machines and self-upgrading individuals as deeply intertwined was ideal.

The person who had been central to facilitating these connections between Shenzhen and American ideals of future making was neither one of the Chinese men in the municipal government nor a foreign maker advocate, but a Chinese woman: Shirley Feng.

Shirley Feng is the director of the Shenzhen Industrial Design Association (SIDA)—an organization that has played a central role in the professionalization of the city's electronics industry; the association connects the thousands of independent industrial design houses at the heart of the region's shanzhai industry both to each other and into a global network of industrial designers. SIDA, in other words, was foundational in building up Shenzhen's reputation as a design city and a hub for industrial design by hosting major international Industrial Design Fairs and connecting Shenzhen's designers to clients and industry partners abroad. Feng came to Shenzhen in 1990. She was a software engineer back then. Her subsequent career became deeply intertwined with the city's transformation and its government's ambitions to position Shenzhen as a hub of high-tech innovation and design—as she told me the story in an official interview she granted me in 2016:

> In 1992, I worked for a car part producer, which originally had done everything by hand, assembling the parts. I built a whole system for them, from import [and] sale to computerization. We used the computer language dBase3 and later dBase+ to set up a database system. I became known in the field because of this early work, and then in 1995, I started working at a pager station, where I became the manager of their technical department. At this pager station, everything worked with computers. It was administrative work. I had to manage only one person; this person did the repair work, and I wrote some small [software] programs, etc. So I was in some sense a technical worker. If a computer was broken, I had to go and buy the spare parts. At that time, the Huaqiangbei markets had just one, single building: SEG. It was just one simple five-storied building. Then, the pager industry was replaced by the mobile phone industry. In 2000, the Internet came, and our pager system was obsolete. I helped transform it into a call center, because the demand for this kind of service was very high—banks, insurance companies, they all needed it. So, we transformed into a call center. The field had one thing: its state-owned companies. The transformation of such companies is not easy. A lot of technologies were replaced too . . . I then set up my own web-development company, called 100 online. At that time, the Internet was so small, so there very few such companies. So, I witnessed the development in Shenzhen from the first building in Huaqiangbei to what it is today. Back then, everybody went to SEG to buy spare parts . . . because of all of this, I have become very familiar with the situation in Shenzhen—with the supply chain, the ecosystem, the creative environment. I think that I witnessed the whole development

of the creative industry, because I came to know all these fields, technical, administrative, etc. And that's why now I am the promoter of this industry.

As Feng explained, her trajectory unfolded through the region's later wave of privatization and industry upgrade, which occurred in the late 1990s and early 2000s.[17] When China entered the WTO in 2001, the development of China's creative industry had been laid out in a series of new policies. While Shanghai had received the lion's share of the official and international attention for the growth of its creative industry (see chapter 2), it was Shenzhen that acted as the laboratory for how the CCP's desired upgrade from "Made in China" to "Created in China" would take place. The focus in Shanghai had been on setting up a new industry, but in Shenzhen the focus was on upgrading and transforming the existing organizational models and industry infrastructures. "If you are working in this industry in Shenzhen," Feng elaborated, "you have to be willing to take risks." In other words, the transformation was not only structural but also ideological—a shift to seeing flexible work and risk-taking as key to the future of the city and China writ large. In 2003, officials from the city government had taken notice of Feng's work: "I had become a 'relatively' famous person by then," Feng told me. Shenzhen's government, eager to transform the city into a "cultural city," invited Feng to help shape the 2003 initiative to associate the city with culture and design. What Feng called "going out" (出去) would play a central role in the success of this project:

> When we started SIDA, we started to think about what an association [like it] could even do. We spent a lot of time thinking about how to do it. I think at that time, I came to a very correct decision: it must be an association with an international perspective. Industrial design just started to really take off, that's why everybody was paying attention. So, if factories in Shenzhen really want to transform themselves, they need industrial design. I was convinced that in the future, industrial design would be very, very important. Very important for our manufacturing, for Shenzhen, actually for China. It would be very important because in China, manufacturing is so important. And in the eyes of foreigners, Chinese manufacturing is considered to be low-level and cheap. It is not a high-level manufacturing. So how can we upgrade from a low level to a high level? Industrial design might be a very important factor. So we have to learn from the advanced countries. Back then, I had the feeling we had to "go out."

The partnership with Gershenfeld and the MIT fab lab promised to be another stepping-stone in what Feng had set out to accomplish in 2003: upgrading the city into a hub for quality design and high-tech innovation, its designers and entrepreneurs were ready to "go out" (just like Robin was aspiring to do) and to change China's image globally and upgrade its national economy.

When I spent time at the various maker and incubator spaces that filled the remodeled parts of the city championed by state officials and foreign maker advocates alike, it became increasingly clear that only certain people were deemed ideal participants in this ongoing project of building China's entrepreneurial factory—investments were made only in those who were seen as prototypic global innovators. In the fall of 2015, I met Zhang Lian on a visit to the newly built Huaqiang makerspace. Zhang Lian had grown up in Shenzhen and went to middle school in Nanshan, "but that was it," he told me, "I couldn't get into high school, because I didn't take the entrance exam. Then I came to Huaqiangbei [in 2008]. There are so many people like me here. People here take classes during work, for re-education. So just like them, I started my own business here." He went on to explain how Huaqiangbei had provided him with a form of training through apprenticeships and learning-by-doing as he immersed himself in Shenzhen's supply chain. Zhang Lian's story conveyed his pride, tied to class consciousness. He explained that "people here in Huaqiangbei were all like this—hard working, continuously learning":

> We are not about theory here. We are about doing . . . People in Huaqiangbei, to be honest, are inferior. They are weak, that's why they need to collaborate and look for partners. So they do well on resource integration. It's not theoretical here. It's all real. We just do things. It's about interpersonal communication and relationships. You can't learn this in books. That's Huaqiangbei culture.

Zhang Lian distinguished all of this Huaqiangbei culture from Shenzhen Bay (深圳湾), the part of Shekou that had been remodeled into the glamorous software park (the venue of the Maker Faire) that I described earlier. Shenzhen Bay was a place where he didn't "yet" fit, Zhang Lian explained: "We understand our position well. We can't go to Shenzhen Bay yet. You can't just go there with nothing. Everything is expensive there, like labor and rent. So you need to think about where you are [positioned in society]. Shenzhen Bay is new. The government and other organizations are supportive of it. I think it's about brand development there, about going IPO, which is risky. When you go there It's a process from turning black to white . . . It's very different from Huaqiangbei. Here, people are emotionally attached to the place. Like me, I don't want to leave. It's almost like a village or a town in the city."

As Huaqiangbei is increasingly imbued with a "maker spirit," and its malls and streets are upgraded to more closely resemble a place like Shenzhen Bay with its globalized aesthetics of tech innovation and cosmopolitan life, it remains unclear where Zhang Lian's place will be. It will depend on his ability to self-upgrade into the kind of entrepreneurial change maker associated with Shenzhen Bay and with the people who had been invited to speak or perform on the stages of Shenzhen's Maker Weeks and Maker Faires.

The Slow Violence of De-shanzhai
(去山寨化 qushanzhaihua)

> In the end, there won't be any shanzhai anymore. But now there is still
> shanzhai. Of course there is! At least within the next 50 years, you will find
> shanzhai in China. This is a slow process of replacing the old with something
> new. This has nothing to do with the industry, with the market. It has to
> do with people. If our kids learned to be a maker, learned about copyright
> protection, learned how to get a patent instead of copying . . . then there really
> won't be any shanzhai anymore.
>
> —TONY, MANAGER, SEG MAKERSPACE, INTERVIEW WITH THE AUTHOR, 2015

This quotation is from a 2015 interview I conducted with the manager of the
newly established SEG makerspace, who went by the English name Tony. Tony,
who held a degree in computer science from the Shanghai Science and Technology University, had come to Shenzhen fifteen years ago when he began working
in the IT department of a logistics company, where he helped set up and keep up
to date the company's software, hardware, and network infrastructure. As Tony
saw it, the key purpose of the SEG makerspace was to enable people to self-
transform by learning from and through making. "The maker spirit can totally
transform the shanzhai ecosystem, because making is about absorbing new things
all the time. This [spirit] is changing the shanzhai ecosystem. There will be patents. Technology, patent, market, network, money. Then there might be a change
in shanzhai." Making would trigger a process of what Tony called "de-shanzhai-
ing (去山寨化 qu shanzhaihua)" and lead to "the creation of brands." This would
take time, he explained. As making infiltrated into education, it would shift the
minds of the younger generation: "Most people don't understand," he explained,
"that maker is about changing the whole structure." All of this would be "a
slow process," and the "key results would only show after the next generation."

Here, Tony was articulating a sentiment that I heard from many people
who had invested in or were otherwise involved in the redesign of Huaqiangbei. Ken, who had founded Huaqiang makerspace, described this same process of structural change through making: "The goal of Huaqiangbei now is
to transform from a grassroots entrepreneurship center into an international
innovation hub . . . Huaqiangbei aspires to develop into the wind turbine of
global hardware innovation. We are committed to building it into the world's
largest experience center, communication platform, and transaction hub in the
fields of IoT and AI." Shanzhai, he explained, was a system that would enable
makers—whom he described as idealists, rather than practitioners—to begin
designing actual solutions and solving problems. When people discussed this
with me, they often implied—and sometimes specifically stated—an assumption that many of these makers would be foreign.

FIGURE 6.7. SEG makerspace—entrance and workspace area. Photograph by the author.

Both SEG makerspace and Huaqiang makerspace had specifically worked to attract foreign makers, partnering with institutions such as the MIT fab lab, European universities, and study abroad programs. Many believed that true and authentic makers—those who were driven by ideals and passion rather than by economic interest—came from abroad. They explained that the Chinese educational system had not been creating such people, and that embedding these foreign makers into Huaqiangbei would embed their "authentic" maker spirit into Chinese processes and value systems. These articulations are reminiscent of the kind of internal orientalism that anthropologist Louisa Schein observes variously expressed during the 1980s opening reforms—the "adoption of Western orientalist logics and premises for self-representation in the course of Asian processes of identity production—processes that are complicit, in their mimetic quality, with universalist, modernizing ideologies."[18] Schein's analysis focuses on images of minority women in the 1980s whose difference signified "both a longing for modernity and the nostalgia that that kind of 'progress' so often inspires."[19] Today, we witness adjacent imaginaries of longing and lack attached to the male shanzhai entrepreneur, whose difference (vis-à-vis international standards of innovation) is portrayed as an "internal other"—on the one hand, viewed nostalgically and as a symbol for what Chinese people have accomplished, and on the other hand, portrayed as outdated and as lacking the skills necessary in China's entrepreneurial factory. Their self-upgrade into globally savvy tech innovators was portrayed as inevitable if they wanted to survive China's latest economic and social transformation.

This was brought home to me when I met the founders of a children's toy design start-up located in Shenzhen Bay, the new software park in Shekou. I had been introduced to founders Wu Ling and Xu Ming through a mutual acquaintance who had moved from Taiwan to Shenzhen in the 1990s to take an engineering and management position at Foxconn. He had recently retired, but he remained well connected to both the more formal and informal economies of electronics production in Shenzhen. Wu Ling and Xu Ming were in their late thirties and early forties, each with children of their own, and were cheerful about showing off their new office. They had just moved here from Huaqiangbei less than two months ago, and they explained how their business model had shifted as they moved locations:

> We did shanzhai five years ago, and now we are all going through a transition. This is a matter of technological development. Or we can say it's the development of history. At first, we don't have the technical know-how or money, so everything is for making money, not design, not your dream. When we have money, we can make better products and bring our brand to a new level. If we fail, we just go back to do shanzhai again. This is how the industry developed. Many companies are following this path. They all start with shanzhai. [. . .] Shanzhai needs a new definition. Is it always equal to copying? I don't think so. In a broad sense, if learning, imitating and reforming a good product is shanzhai, the term would be positive. It is a learning process. Any country or industry needs to go through this phase to develop. You can say that shanzhai is a way to keep a company or factory alive in the beginning. There is no other way around it. It's not like people want to copy. That's not the right attitude to do business or products. But there is no other choice. . . . If you refuse to do it, you can't survive. That's the process of development.

With their transition from Huaqiangbei to Shekou and from shanzhai to what they considered Western-style making, the two co-founders also began to shift their target markets from Chinese to foreign end-consumers. Being in transition meant moving from shanzhai into brand development and the land of intellectual property: "When a company is giving up its shanzhai identity, it needs to move out of Huaqiangbei and start to do middle-end or high-end products. They won't focus on the Huaqiangbei market anymore." Their professional identity had evolved through and alongside shanzhai, specifically the shanzhai mobile phone industry in Huaqiangbei. One of them had been focused on developing software solutions around the MTK chip, which he claimed had "made him tons of money," while the other had been working on hardware design. Neither of them had a college degree; both had followed an apprenticeship model. They both started their careers working for tech corporations in Guangzhou. In 2003, one of them moved to Shenzhen, with the other following a year later to start their own businesses.

Wu Ling and Xu Ming had entered the shanzhai industry at a time when it was thriving. Over my years of research, I met many entrepreneurs like them; people who had recently left Huaqiangbei in order to shed their past affiliations with shanzhai. They emphasized the value of the knowledge they had gained through shanzhai, which many described to me as having an in-depth understanding of 产业链 (chanyelian—the whole supply chain or the complete production chain). Most of the people with more than ten years of professional work experience in Huaqiangbei had acquired a largely tacit knowledge of the different entities that made Shenzhen's electronics production tick, from assemblies to component producers, molding factories, stamping factories, design solution houses, prototyping houses, packaging, vendors, and many more—thousands of corporations of varying sizes and varying degrees of reputation and quality. These people had retained an inner map of the key players that made up the supply chain, of how the scene had shifted, and of who was a reliable partner. This know-how was concentrated in individuals or in individual corporations—and while rendered as in need of an upgrade, was also portrayed as an asset for survival.

A company that had become known—in the shanzhai circles where I was conducting my research—as having successfully managed this self-transformation was TopDesign,[20] a prominent design house in Shenzhen. TopDesign had become known as a place where up-and-coming designers and engineers trained, many of whom would then later set up their own competing businesses in the local industry. "The foundation of Shenzhen was people struggling together, building networks together, making money together. This is the foundation of Shenzhen and this is what is specific to Shenzhen," William,[21] one of TopDesign's lead engineers, explained this to me.

I got to spend more time with William on a February 2016 tour, "Shenzhen Makers in Silicon Valley," which was funded by the Shenzhen international exchange foundation and the Shenzhen government bureau for foreign affairs. David Li, in his capacity as director of the Shenzhen Open Innovation Lab, had been charged with organizing the three-week-long trip to the Bay Area and leading the group, and he had invited me to join. The tour was meant to introduce ten Shenzhen-based designers, engineers, educators, and artists—whom the sponsors of the tour referred to as "Shenzhen makers"—to Silicon Valley innovation by visiting a variety of organizations and companies, from large tech giants such as Intel and Google to smaller start-ups, incubators, maker- and hackerspaces, and coworking spaces, as well as university spaces such as Stanford's D-School and Production Realization Lab. The visit to Silicon Valley struck me as a parallel to the trips to Shenzhen made by Western makers and open source advocates I describe in chapter 3. While the Western makers had construed their forays into Shenzhen as a "going back in time" to learn from the "fearless shanzhai experimenters" how to hack economies of scale and capital's speed, the Shenzhen makers understood their visit to Silicon Valley as learning the reverse: slowness and making for fun. "There are no real makers in Shenzhen," William of Top-Design told me after his visit to the Bay Area; "people here [in Shenzhen] care so much about business and about making money. Chinese people are not used to sharing something in an open space. Chinese people are not good at that." I came across similar self-essentializing and internal orientalism from people on numerous occasions. It was a central theme during the Shenzhen maker group's visit to Silicon Valley, with many of the Shenzhen-based engineers and designers insisting that the form of making that they saw there was the "true" spirit of open source and technology innovation, which—they argued—China lacked.

One of the stops on the Shenzhen makers' tour to the bay area was the accelerator TechNode, which hosted an evening event where each of the visitors from Shenzhen had a three-minute time slot to "pitch" themselves and their work. The morning of the event, the anthropologist Lyn Jeffery hosted the group at the Institute for the Future (IFTF), where she organized a session for the Shenzhen visitors to practice their three-minute pitches: "three-minute pitches are very difficult," Jeffery explained to the group gathered in a conference room of the institute, "but it's something that is common practice here." Jeffery has been researching entrepreneurship and tech cultures in China since the 1990s, and she has been an important voice dispelling the associations between Chineseness and a lack of creativity, both in China and in Silicon Valley. She framed the practice pitch session as helping the group from Shenzhen articulate and communicate clearly why their work was just as creative and innovative as what people in Silicon Valley were doing.

Visits to Silicon Valley by Chinese tourists, businesses, and governments have become an industry of their own, mirroring Chinese tours to European cities like Paris and London. This form of cultural travel is largely aimed at cultivating in Chinese people a supposedly absent attitude of cosmopolitan consumption and civility. The government entity that had sponsored the Shenzhen makers tour to the bay area was aiming for something similar; in many ways ironic, the aim was to cultivate a supposedly absent spirit of innovation and entrepreneurial capacity (the kind associated with start-up pitches, incubators, and makerspaces). But David Li was ardent that the purpose of the trip—in his view—was quite different: one of mutual learning and of challenging the very notion that Silicon Valley had nothing to learn from Shenzhen. He repeatedly reminded the tour attendees that "People in Silicon Valley should recognize how much they can learn from all of you." He saw Lyn Jeffery and myself as allies of sorts in this project—as ethnographers, each of us had been committed to challenging the way that the non-West was undermined and othered by the American and European design and engineering industries. Throughout the trip, debates—at times heated—took place between Li and the "Shenzhen makers" around this question of China's position with respect to Silicon Valley. The Shenzhen makers tended to be humble in their three-minute speeches, downplaying their past accomplishments and highlighting their ambitions to bring a maker and innovation spirit to China by learning from Silicon Valley. "I want to know how the shanzhai and maker movement can come together. I want to show people that we are changing from shanzhai to something else," one of the engineers ended his pitch. "Don't say that!" Li quickly responded. "We actually don't want to tell this audience that shanzhai is a bad thing." Others related their pitches to their experiences in Silicon Valley, stating that they aspired "to become a maker just like the people [they had met] at the Stanford D-School" so that they could enable other people in China "to acquire creative thinking" and "accelerate the development of makers." Li intervened frequently, attempting to get them to see their own work as a remarkable, innovative accomplishment.

For Li, the tour participants' humble attitude and eagerness to learn proved that shanzhai and Chinese culture writ large did indeed constitute an alternative to the culture of masculinity and "brogrammer" elitism embodied by Silicon Valley. As Li saw it, their survival tactics, their ability to make out of necessity, and the consumer-oriented (rather than venture capital–oriented) value of their businesses meant that they were makers who were actually implementing the promise of making—who were actually working toward the maker ideal of a more egalitarian future for technology production and innovation.

As the Shenzhen makers stood, somewhat awkwardly, at the front of the IFTF conference room to deliver their pitches, I was reminded of the many pitch practices I had witnessed by Western start-ups and maker advocates in China. It suddenly dawned on me how odd it was that the makers from Shenzhen were expected to deliver their content in English. None of the foreign start-ups or makers I had met in China were ever expected to give talks in Mandarin Chinese. On the contrary, much effort had been put into supporting the "foreign guests" by providing Chinese-English translators and interpreters, and often, if these resources weren't available, the default language was English.

When I returned to Shenzhen later that year, I again met some of the people who had participated in the tour to Silicon Valley. William had offered to show me his company, TopDesign. When I arrived at the design firm, William and his colleague Jenny,[22] the firm's director of PR and marketing, welcomed me in the lobby and took me on a tour. I saw the large showroom, the in-house prototyping and production facilities, the design studio, the artifact library, the engineering lab, and the TopDesign academy. The impressive spaces showcased the company's history and how it has transformed since it was founded in 1999. In the showroom, Jenny told the story of the two co-founders, each of whom had worked in large design houses in Hong Kong and had given up lucrative salaries to move to Shenzhen and start a company of their own. What allowed the company to succeed, Jenny and William explained, was that its co-founders not only understood the region's supply chain but had designed their company to map and eventually shape that supply chain. TopDesign bought a molding company in 2002 and became one of the first design houses in the region with its own integrated prototyping department—a function that most other companies outsourced. The company made its workflow and the method public; both can still be accessed today on the company's website.

The company had begun by working in the shanzhai mobile phone industry. TopDesign's showroom not only included the company's current wide variety of product lines, designed for well-known international firms in the car industry, telecommunications, entertainment, household electronics, and medical equipment: Philips, Disney, Haier, Huawei, Midea, Whirlpool, ZTE, Verizon, Audi, Bentley, Panasonic, Acer, and more. It also had on display a series of feature phones that dated from the company's early days—devices that I had seen and purchased versions of in the Huaqiangbei electronic markets. TopDesign, it appeared, had been a trendsetter for the design of many of these phones; TopDesign had produced reference designs for the shanzhai industry, shaping the aesthetics of shanzhai production for many years to come. In local industry terms, this practice of modeling after leading design firms was often referred to as *gongmo* (public shell/exterior). When I began my research in Shenzhen in early 2013 and started collecting shanzhai phones, I wanted to know how shanzhai worked—its material production and design practices. While much has been written about shanzhai (as described in chapter 3), there was relatively little research and writing about how shanzhai's so-called open sharing culture worked in practice and what motivated it. Who produced phones shaped as a Hello Kitty figurine and who carefully designed golden Buddha phones as an out-of-the-box experience? If shanzhai entrepreneurs could make a phone within a month (as more and more media outlets, open source hardware advocates, and scholars claimed), how was this actually done?

As I explained in chapter 3, at the heart of shanzhai production were two key artifacts: gongmo (public face or shell) and gongban (public board). While gongmo referred to the encasings and the aesthetics of the devices, gongban referred to the interior—the printed circuit board. Companies like Apple are known for keeping both aspects as closed (proprietary or private) as possible. In the shanzhai industry, both are understood as public. Gongban can also be translated as "reference board," a layout and design of the board, including a list and arrangement of its components. Such reference boards are commonly produced by chip manufacturers in order to recommend a board design that would enhance the functionality of their chip technology. The Taiwanese chip manufacturer MTK, which had played a central role in the formation of shanzhai, publicly released such reference designs, as did a number of distributors whose main business was focused on selling components to assemblies and design solution houses. In 2013, I met one of the region's largest distributors in the industry. The distributor had its own engineering division which produced roughly 130 different gongban per year for a variety of devices, from mobile phones, tablets, and smart watches to washing machines, industrial robots, and vehicles. The firm offered gongban designs based on Intel chip technology as well as reference boards that were based on more affordable chip sets, including MTK, Rockchip, Qualcomm, and others. The company would not sell these boards; as a distributor, it was in the business of trading components between the component producers (e.g., Intel, MTK) and the local assemblies and design houses. The open reference designs assured these factories that the components were compatible, so keeping reference designs open was in their businesses' interests. The distributor also made recommendations about which "skins" or "shells" were compatible with the boards. This basically meant that distributors recommended design houses within their own professional networks— companies like TopDesign. Although no public molds or drawings were released for gongmo designs, it was easy and affordable to reverse engineer these molds— much easier than reverse engineering the electronics. Many small molding and design houses specialized in exactly such processes. TopDesign's designs were among those widely copied, often with only minor alterations.

In TopDesign's showroom, the company's early phones were displayed right at the entry, in carefully arranged glass displays. As I wandered through the aisles, I noticed that one case stood apart from the others, spotlighting one particular phone in greater detail. There it was on display—my Buddha phone, the highlight of my shanzhai phone collection as I described it earlier! The label on the case read in English "The Buddhism Mobile Phone" and in Chinese 禅机 (chanji—chan is the character for deep meditation/Buddhist and ji is the character for machine, as in shouji 手机, which means mobile phone). The label included the following short description: "This phone is inspired by the Buddha Pendant that has blessing spirit that 'water nurtured all things,' showing culture of Buddhist as the core, promoting the crafts and human wisdom to unprecedented peak. It is made of millennium dazzling jade and pavilions beautiful gold." Jenny explained that one of the co-founders of TopDesign was Buddhist. In 2010, he and a good friend sat over tea lamenting the irony in the fact that they could design great technologies for other people but not for their own religious practice. Would it be possible, they wondered, to design a phone that supported rather than disrupted some of the key principles of their religion? Thus the idea of the Buddha phone was born. TopDesign released only a small batch of a few hundred pieces—this phone was not meant to make a profit. It came in three editions: "platinum," "reserve," and "collectors," with prices ranging from RMB 3600 to RMB 9600.[23] Just like my phone, all the Buddha phones came with a built-in jade stone, a wooden box with golden silk linings, a Buddhist collectable card, and carefully designed accessories.

FIGURE 6.8. TopDesign's design reference library. Photograph by the author.

The only apparent difference was the price—the phone I had purchased in 2013 had only cost me RMB 400. Had I acquired a derivative of the TopDesign gongmo? When I came home several weeks later, I compared the photos I took in TopDesign's showroom with my own phone. I couldn't find a single difference. The jade stone, the accessories, even the little Buddhist prayer card—it all looked exactly the same. There was even a certificate for the jade stone in my box. Again and again I looked at the photo I took in TopDesign's showroom and the phone in my hands. As I was about to give up, I noticed that the wooden box—while identical in its design—carried a different label. While the TopDesign box carried the name "wellwishing," mine read "sky good luck" and instead of 禅机 it read 佛教用品 (product for Buddhism). I looked again at the phone and began noticing subtle differences in the layout of the keyboard and a different rounding of the graphic's corners on the back of the phone. It was a derivative.

My Buddha phone could be seen either as a cheap knockoff of TopDesign's design or as embodying the spirit of shanzhai that so many—especially Western commentators—had found remarkable (see chapter 3); what would have remained out of reach, at a price of several thousand RMB, had been made accessible for many more people. If one adopted the former position, the Buddha phone in my possession was a cheat, deceiving consumers into purchasing

FIGURE 6.9. "Buddha Phone" in the TopDesign showroom. Photograph by the author.

a device of low quality. If one took the latter position, the careful and detailed act of replicating could be seen as an art form in its own right.

Whereas for many of the Western maker advocates who have flocked to Shenzhen since around 2010 shanzhai promised to recuperate the ethical commitments of open source and technological promise itself (see chapter 3), William and the other Shenzhen makers who had joined the tour to Silicon Valley that year felt highly ambivalent toward my shanzhai Buddha phone; on the one hand, it was seen as a shameful aspect of China's past (its associations with economic greed, low quality, and fake). On the other hand, it was viewed with a sense of nostalgia and a sense of pride for what Shenzhen and its industry had accomplished: the buildup of a global supply chain in electronics production and design without which Western tech innovation would have been impossible.

The hope was that if Shenzhen's designers and engineers transitioned from shanzhai to branded goods, from Huaqiangbei to Shenzhen Bay, from economic to entrepreneurial actor, from taking risks by coming to Shenzhen (from another part of China) to taking risks by "going out" (becoming a globally recognized Chinese tech entrepreneur), they would also demonstrate an ability to move toward the future—they would be recognized as innovators by the people at Stanford's D-School, at the corporate offices of Intel and Google, and perhaps most important—they would be able to gain a sense of self-confidence and national accomplishment, freed from connotations of fake, low quality, and "other."

Mandatory Happiness

Although many of the people I met over the years of this research told me that they thought Xi Jinping's promise to build a "Glorious China" (辉煌中国 huihuang zhongguo) was good for China,[24] they also agreed that it was not necessarily good for them personally. In particular, Xi's invocation of the Chinese Dream (中国梦 zhongguo meng) often evoked mixed feelings in the harsh reality of China's entrepreneurial factory. People spoke with me about the precarity of their situations and the constant, inevitable, exhausting need to yet again re-fashion themselves. They experienced, in other words, what is often described as a symptom of neoliberal capitalism; when individuals become human capital, life is reduced to a limited form of human existence that is concerned only with survival and wealth acquisition.[25] China's entrepreneurial factory occludes its own violence (the precarious conditions of work it demands and that operate alongside the internalization of Western colonial and racialized othering as I described it in chapters 2–5) by promising participation and agency—it thus occludes how happiness became a mandate, a mechanism of state control. This was brought home to me in 2016 in an encounter with a Chinese taxi driver. Ingrid Fischer-Schreiber, who was traveling with me at the time, and I had just left a meeting with a young designer in his mid-twenties, who had moved from Hong Kong to Shenzhen a year ago to start a company with four former classmates and colleagues. The taxi was old and dirty, and reeked of a mixture of gasoline, polluted air, sweat, and takeout food. It could not have been more of a contrast with the design studio we had just visited, located in one of Shenzhen's new innovation parks. Stepping out of the modern building, with its glass surfaces and interiors equipped with Eames furniture, and stepping into the old taxi felt like entering an older China, one that new China was ready to discard. "Where are you from?" the driver asked as the car began to move. "What do you think of China?" Ingrid and I explained that we liked China and that we had lived here for many years—the kind of thing I often repeated when curious strangers asked me about my language skills or why I was in the country. The taxi driver interrupted our explanations and declared firmly, "China is not good. China is bad. The government is bad. The government is rotten. They don't care about us. They don't care if we survive or not. They just care about themselves." The driver continued his complaints for the rest of the twenty-five-minute ride, angrily telling us how the government endorsed new models of economic development without any regard for what these changes meant for its people, shouting that the government encouraged digital ride-sharing platforms such as Didi Chuxing to enter the market without attempting to protect existing industries. Chinese taxi drivers were often a font of knowledge about social and economic shifts in China, and I had often debated politically sensitive matters with them over the five years I lived in China. But I had never encountered the rage this driver expressed. "The government is rotten," he stated again and again. We listened and nodded.

I was reminded of the Shenzhen taxi driver as I read the recent news story of the NYC taxi driver's suicide. His last words, posted on Facebook, articulated his despair, anger, and numbness—the feelings that Chinese taxi driver had articulated: "Companies do not care how they abuse us just so the executives get their bonuses. Due to the huge numbers of cars available with desperate drivers trying to feed their families, they squeeze rates to below operating costs and force professionals like me out of business. They count their money, and we are driven down into the streets we drive, becoming homeless and hungry. I will not be a slave working for chump change. I would rather be dead."[26]

While precarious work in China began spreading in the late 1980s and was accelerated in the 1990s when the socialist system of the iron rice bowl was replaced with a flexible labor model,[27] the Shenzhen taxi driver spoke to a more recent variation. The global uptake of digital labor platforms has worsened global division of labor, affecting the living conditions of those who were already excluded from more stable job conditions; it has also damaged jobs previously considered fairly stable, white-collar and middle-class, from taxi driving to translation services to design. These jobs have been transformed into a system of micro-payments and tasks that, in the short term, lowers prices of services for consumers (taxi rides) but ultimately increases the profits for the tech corporations that own the platforms.[28] These digital labor platforms continue old forms of labor exploitation, spreading slave-like working conditions as their infrastructures instantiate the continuous feminization of work.[29]

Displacements of technological promise via the entrepreneurial factory further legitimize exploitation by concealing it behind the promise of the good life and justice. Moments of anger—like the one I had experienced in the taxi—are rare in China's entrepreneurial factory. When the good life appears to be within reach, the political and corporate coercion of happiness seem like "innovative" approaches to business and politics—this is the slow violence of technological displacement.

Only a few months after the CCP's official announcement of the "mass makerspace" (众创空间 zhongchuang kongjian) initiative in January 2015 (see chapter 1), China saw thousands of such spaces popping up. I visited several of these spaces and met with people who had been involved in orchestrating or funding them. Oftentimes, these spaces were added in preexisting creative industry or innovation parks that had been built in the wave of real estate development triggered by the "creative industry" policies of the early to mid-2000s (see chapter 2). I met no one who had received direct funding from the central government to establish these spaces. Instead, preexisting spaces were equipped with low-end 3D printing machines, laser cutters, soldering irons, working tables, and colorful chairs. The people who ran these spaces hoped that they would qualify as official "mass makerspaces," which in turn would mean the possibility of investment from district- or municipal-level government entities that had been charged by the central government with implementing the CCP's "mass maker-space" initiative. Investors and low-level government officials told me that they hoped that these "mass makerspaces" would attract business to their districts, and that the resulting stronger economy would help fill China's innovation parks and creative industry clusters, many of which had remained empty.[30]

These "mass makerspaces" were met with ambivalence by many of the people I had come to know over the years of my research. While the CCP's maker initiative itself was seen as aiming at the kind of change many had long desired (to establish China as a nation of innovators), many saw the sudden rush to open "mass makerspaces" nationwide as motivated by the desire to make quick money. While people critiqued the low-level politicians and real estate businesses who attempted to make money from the initiative, the CCP itself was understood as supporting people's long held desires to upgrade or altogether replace China's "humiliating" shanzhai businesses with "innovative" companies that carried international stature. The CCP's appropriation of making seemed like a good thing, even hopeful. The 2015 mass makerspace policy and the urban upgrades (and the forms of slow violence they legitimized) were interpreted as a sign of the government's commitment to supporting people's own desires to achieve parity with the West—the CCP was seen as serving citizens' interests to achieve happiness.

The character for "mass" (众 zhong) in "mass makerspace" invokes China's history of grassroots mobilization that was central to China's political system under socialism in the twentieth century.[31] When the CCP announced the mass makerspace initiative, however, it did not advocate for a return to the kind of revolutionary uprising or political mass mobilization of that era. On the contrary, grassroots mobilization served the function of the pitch; the ultimate purpose of the mass makerspace initiative was to create an affective bond of mutual interest between the CCP and China's citizens—to create the feeling that the CCP was an ally (a peer) acting in the interest of the people as it was prototyping China anew—as a nation of dreamers and happy innovators—both nationally and globally.

When I saw Robin last, he sent me a WeChat message with a news clipping from the *Daily Mail* in the UK. "Shenzhen, China's reform pioneer, leads tech revolution" read the article's headline. Not without a hint of pride, Robin directed me to read till the end, where the article featured him *right next* to a white man—the managing director of a well-known Shenzhen-based incubator program. It is this promise of happiness—as Sara Ahmed so insightfully points out—granted via attachments to whiteness that made many people I met in China's shanzhai industry endure the slow violence of the entrepreneurial factory and its demand to take more and more risks, to self-economize, to overwork, and to *take responsibility* on behalf of the nation.

FIGURE 6.10. Mingtong Market in Huaqiangbei, before renovation. Photograph by the author.

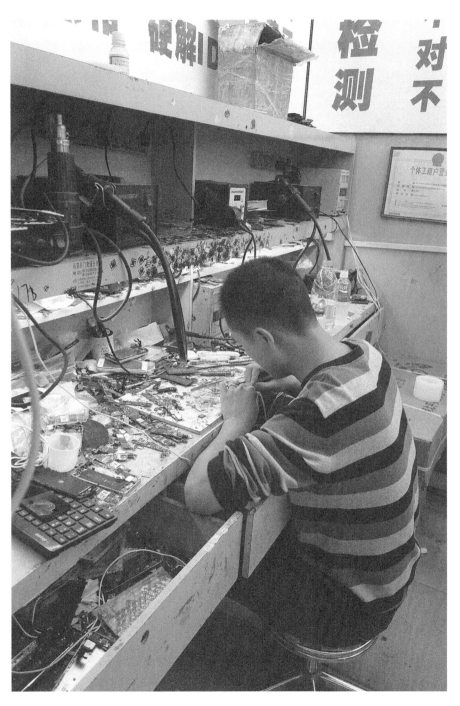

FIGURE 6.11. Repair shop in Huaqiangbei, before renovation. Photograph by the author.

FIGURE 6.12. Mingtong Market in Huaqiangbei, under construction (left side upgraded, right side renovated). Photogrphs by the author.

FIGURE 6.13. "Chinese Dream" campaign by the government on wall in front of construction site in Huaqiangbei. Photograph by the author.

FIGURE 6.14. Banner in front of construction in Huaqiangbei promotes "International Maker Center." Photograph by the author.

FIGURE 6.15. Store in Huaqiangbei, after renovation. Photograph by the author.

FIGURE 6.16. Store in Huaqiangbei, after renovation. Photograph by the author.

FIGURE 6.17. Huaqiangbei, people walk leisurely at night, after renovation. Photograph by the author.

FIGURE 6.18. Huaqiangbei, after renovation; signage in English and Chinese "Huaqiangbei—The First Electronic Street of China" and "华强北—中国电子第一街." Photograph by the author.

7

Conclusion

THE NURTURE OF ENTREPRENEURIAL LIFE

> If we end the story with decay, we abandon all hope—or turn our attention to
> other sites of promise and ruin, promise and ruin.
> —ANNA TSING, *THE MUSHROOM AT THE END OF THE WORLD*, P. 18

> The silly or ridiculous nature of alternatives teaches us not about the nature
> of those alternatives but about just how threatening it can be to imagine
> alternatives to a system that survives by grounding itself in inevitability.
> —SARA AHMED, *THE PROMISE OF HAPPINESS*, P. 165

Displacements of technological promise do not simply happen due to some
inert law of capitalism or due to some "natural" functioning of technology
itself. Rather, the promise of technological intervention, of progress, and hopeful future making endures, because it is nurtured and incubated. As I have
shown in this book, displacements of technological promise—attachments of
technological promise to sites and bodies once labeled incapable of innovating (from rural America to Chinese factories)—depend on happiness labor,
on the deliberate design of workspaces, select neighborhoods and zones to
induce the feeling of experimental agency, and on training and incubation.
In chapters 4 and 5, for instance, I have shown how incubators train and cultivate human capital by pitching self-economization as democratized peer
production and carrying interventionist capacity (economic life is celebrated
as entrepreneurial life); and I have detailed the happiness labor necessary to
sustain the socialist pitch of entrepreneurial living that venture capital needs

for its own survival. In chapters 3 and 6, I have chronicled how the material and discursive refashioning of certain spaces, practices, and bodies as "other" and "outside" the modern innovation economy (such as China's industry of shanzhai manufacturing and design) assigned them interventionist agency, making them attractive for capital investment. The promise of making—that the democratization of technological experimentation would provide the tools necessary to intervene in established structures of corporate control and exploitative digital technology, and create alternatives outside techno-capitalistic greed—rested on those formerly excluded from creative tech production (people denied the promise of happiness attached to tech innovation from women of color hired as happiness workers to the designers and engineers working in China's shanzhai industry). The ideals and practices of a global maker movement *and* the imaginary of a "new" China (understood as providing the "toolkit" necessary to prototype and intervene at scale) were co-produced amidst a broadening suspicion of the tech industry, research, and education, their attachments to venture capital and complicity in exploitation and in channeling resistance and intervention for the purposes of capital value accumulation for elites. The tenacity of technological promise in this very moment of a broadening critique of the tech industry depended on the select recruitment of once excluded bodies and sites.

Through the displacements of technological promise, certain forms of life are subsumed under capital production, while others are continuously excluded. As the economist Kalyan Sanyal shows, capitalism operates not only through subsumption but also exclusion: "Capital's strength lies in its ability not to annihilate its others, but to negotiate the world of difference." Capital, Sanyal elucidates, produces "non-capital"; in other words, the exclusions of certain bodies and sites from the sphere of capitalist production—their trapping outside the circuit of capital—is part and parcel of its workings. It is the promise of participation in entrepreneurial life that in turn legitimizes such exclusions as a temporary glitch, to be overcome if one learns how to innovate "properly." It is important to understand that displacements of technological promise are *not* a linear movement of technology and capital investment from "here" (the so-called developed parts of the world) to "there" (the so-called under- or less developed parts of the world); they unfold, instead, through recursive motion, reappropriations and recuperations of certain pasts, of select sites and bodies that promise to "pitch" well, *and* through the exclusion and silencing of certain pasts, bodies, and sites. I have shown in this book how the making of Shenzhen as a new "frontier" of technological innovation, for instance, was co-produced by the endurance of colonial tropes of China as "other" and "backwards," by a proliferating distrust of (Western) modernist ideals of progress (which found expression in a growing number of attacks launched at the tech industry and Silicon valley in particular) *and* the CCP's project to (re)assert China as a global empire—a project that deliberately reappropriates China's partial colonization in the nineteenth and early twentieth centuries.

While displacements of technological promise took specific form in contemporary China precisely because of these tangles of past and present, of colonial endurance and the mobilization of desires, they have their countless parallels. To name but one recent example: in December 2017, the Republican Party introduced the *American Tax Cuts and Job Act*. The initiative included a so-called opportunity zone program, aimed at funneling private investment into "economically distressed" neighborhoods in exchange for a tax break on the capital gains.[1] According to a *New York Times* article from August 2019, there are now more than 8,700 such zones in existence nationwide.[2] The opportunity zone initiative has not only drawn developers and capital investors, but also incubator and coworking spaces[3]—organizations that hope to benefit from the opportunity zone's proximity to capital and that in turn appeal to investors, for they promise to move innovation talent into the zone. In other words, exactly because they were previously excluded from the circuits of venture labor[4] and venture capital, neighborhoods are portrayed—quite literally—as promising, as a unique opportunity and an asset for the investor. Neighborhoods, regions, bodies, sites, zones once excluded compete for investment by "pitching" their specific social and economic inequality and poverty as attractive to the anticipatory logics of finance capitalism. Resistance to investment programs that deploy this socialist pitch—the economization of life "pitched" as enabling social and economic change—is reduced precisely because the programs *appear* aligned with commitments to justice.[5] The socialist pitch derives its power from producing an affect of interventionist capacity and change. By creating a feeling that change is possible, the socialist pitch channels hopes and yearnings for alternatives into the demand of fashioning the self as an entrepreneurial agent, who approaches the economization of life with exuberant excitement and hope.[6]

This book examined the ways in which this seemingly inevitable spiral of exclusion and promise, displacement and anticipation, yearnings for alternatives and their subsumption has to be nurtured, sustained, maintained, and actively cultivated—to show how entrepreneurial life has to be nurtured. If we attend to the bodies, places, instruments, and techniques that are necessary to cultivate and maintain entrepreneurial life, we can begin to think otherwise about the relationship between technology and markets; we notice that the transformation of economic life into entrepreneurial life—the imbuing of economization with feelings of interventionist capacity and agency—is "made" rather than following a technologically determined script. The study of the labor, the spatial and infrastructural design, and the training programs that are necessary for technological promise to endure, I argue, can challenge economization's apparent fait accompli and make visible the violence that the socialist pitch of entrepreneurial life occludes. Laboratories of experimentation—from makerspaces and fab labs to the Special Economic Zone and opportunity zone—produce feelings and hopes for change, but not necessarily change itself. In fact, it is the *feeling* of intervention and agency they produce that masks their violence—racism and colonial othering, sexism, gender discrimination, old and new forms of labor exploitation, and the spread of precarious conditions of work and life. In the political and economic regime of technology innovation and venture capital, capitalist expansion is pitched as participatory—available to "everyone." Processes of economization and capitalist expansion no longer appear to be just a "natural consequence" of innovation and competition, but rather appear to be "hackable," something we can make our own. In other words, the neoliberal doctrine of inevitability (the idea that automation, AI, capitalist expansion, and so on are natural, inevitable processes rather than the outcome of human choices) is imbued with a feeling of agency—economic life is framed as entrepreneurial life.

But just as the "nature" of the capitalist market—the idea that markets self-regulate and operate based on some inert quality of capital itself—has to be actively produced and continuously maintained (via market devices, including ironically state intervention), so do the markets of finance capital and speculation have to be nurtured (via the production of feeling). While market devices such as pricing techniques, accounting, monitoring instruments, trading protocols, and the training of economists and traders translate complex societal processes into quantitative measures and seemingly controllable entities, the market devices of venture capital such as the socialist pitch of entrepreneurial life channel yearnings for justice to create feelings and anticipation of change. The socialist pitch and the production of hopeful anticipation depend on happiness labor—the labor that produces a feeling of optimism and hope despite the proliferating sense of rising precarity. If we attend to the labor and the instruments of affect that finance capitalism needs, we notice the vulnerability of capitalist production—that is, we notice that the relationship between technology, life, and markets can be otherwise.

Noticing Complicities

I am making the final edits to this book amidst the heated controversy over the MIT Media Lab's involvement with and acceptance of donations by the financier and convicted sex offender Jeffrey Epstein. Joi Ito, the former director of the MIT Media Lab, resigned a day after *The New Yorker* published an explosive essay by Ronan Farrow that revealed Ito's attempts to conceal contact with Epstein and the depth of their fund-raising relationship. When I began the research for this book ten years ago, a scandal like this was unthinkable.

Ito—alongside the men who had come to his rescue after the connections to Epstein were first revealed (among them Lawrence Lessig, whom I have also discussed in this book)[7]—was admired and celebrated as an advocate of the free culture movement committed to intervening in the intellectual property regime and corporate control by prototyping alternative structures of creative tech production (as embodied, for instance, in Creative Commons). When Ito was appointed as director of the Media Lab, many felt hopeful. Ito had positioned himself as an "outsider" to the bureaucratic structures of higher education and—via his involvement with Lessig's Creative Commons—as an advocate of the kind of piracy that Lessig had labeled as moral and good and as in contrast to Asian piracy, which he argued was "bad."[8] Ito's appointment thus created a feeling that change and intervention into elite educational structures was possible (see chapter 2). When Ito, alongside other prominent free culture and open source advocates, turned to Shenzhen in 2014, the region became imbued with this very feeling that justice and change were possible if one prototyped at the scale of bureaucratic structures, markets, and global supply chains.[9] The idea that Shenzhen was an emerging frontier of innovation exactly because it supposedly stood for "bad" piracy, for capitalism on steroids, and modernity's "other" (advanced not only by Ito but by several prominent figures of the tech industry and investment) was widely picked up and celebrated by media and scholars alike for its intervention into Western-centric notions of design and innovation rather than for the colonial othering it also was (see details in chapter 3). Or, stated differently: it was difficult to notice the masculine and colonial tropes of adventure and control that co-produced Shenzhen as an emerging frontier of innovation, exactly because they were embedded in a feeling of radical change and the promise of justice and interventionist capacity.

It is yet to be seen if the revelations of the MIT Media Lab's involvement with Epstein will lead to a broader reckoning of how universities have been restructured via the socialist pitch of entrepreneurial life. Entrepreneurial life is pitched as providing students, faculty, and staff, and by extension the university, with a renewed capacity to make "actual" change (rather than "just" knowledge). As I have shown in chapter 4, in part fueled by debates over the societal and economic consequences of student debt (debates that increased after the 2007–2008 financial crisis), the university was increasingly construed as the antipode to entrepreneurial life; bureaucratic, slow, hierarchical, top-down, incapable of change and enabling change. The story was that the skills students acquired at the university were outdated, incapable of preparing them for the real world. This image of the university as lacking and lagging was further fueled by the claim—ironically brought forward by scholars themselves—that critique itself had outlived its usefulness, that it was negative, and that academia—and the humanities in particular—were always just a step behind, catching up with the phenomenon under study—as STS scholar Bruno Latour notoriously put it—"one war late, one critique late."[10] These claims made from both outside and within the academy continue to legitimize the restructuring of higher education. University incubators, fab labs, and makerspaces, the use of "design thinking" toolkits in the classroom, and the hosting of hackathons at career fairs all promise to bring the university "up to speed"—they teach students how to build prototypes of intervention and pitch the university as a portfolio piece that promises to help the student "stand out." The add-on of these "experimental" spaces unfolds alongside the shifts in allocation of resources within the university and outside the university, with funding channeled toward disciplines and institutions that promise high enrollment and high return on investment. The better a university or a department "pitches" itself as *the* producer of change makers, the more lucrative its assets.

While the Media Lab might be one of the prototypes of this process of shaping education according to the interests of investment, it is the many sites that have variously taken up the socialist pitch of entrepreneurial life that further sustain it. At the same time as a range of governments, the American military, venture capitalists, and philanthropists began investing in making, it also attracted the attention of researchers and designers who had long been critical of the political and economic regimes of technology production and design. It was taken up, for instance, by researchers in the field of participatory design in a 2014 edited volume that stipulates fab labs and makerspaces as building on and "scaling" PD's early political commitments to broaden participation in design (via the mobilization of both technology *and* labor movements)—as "platforms for broader participation and new ways of collaborative engagement in design and innovation, pointing at alternative forms of user-driven production." Postcolonial scholar Arturo Escobar builds on this argument, speculating that fab labs and makerspaces demonstrate how design—"liberated from the imagination that the world of the south has in large part been an ontological designing consequence of the Eurocentric world of the North"—can become a 'toolkit' for reimagining and reconstructing local worlds."[11] In a similar vein, researchers and practitioners in the field of human-computer interaction have theorized making as providing a toolkit to center feminized practices of tech production (e.g., craft) as such opening up design and computing. In adjacent fields of science and technology studies and critical computing, making was theorized as empowering people to define civic life on their own terms (independent of the state or other powerful actors) and as recuperating critique itself via hands-on tinkering and experimentation.

This turn toward making by scholars from a range of fields was shaped by and also fueled the claim that the familiar tools of both critique and intervention that the scholar had drawn from were ill-suited to address a seemingly all-subsuming capitalism, the social factory, and its proliferation of precarious conditions of work and life.

In this book, I have used the prototype as an analytical concept to think through the shifting imagination of technological innovation. Much of the continuing allure of technology production and design stems from its early associations with "tinkering" and "experimentation," and the enticing promise that *technological prototyping* would serve *societal prototyping*. When the early prototypes of the personal computer were introduced, they demo-ed not only technological innovation but also that digital technology was an instrument, placed in the hands of individuals, to prototype alternatives—to alter people's view of the world by testing out and experimenting with alternatives. When Douglas Engelbart in 1968 performed at the Association for Computing Machinery what would become known as "the mother of all demos," for instance, he demo-ed not merely the fundamental elements of personal computing, but also the idea of an empowered technology "user" who manipulated digital structures and content with a few mouse clicks. Engelbart had prototyped and "modeled" a different way of looking at the world—a world that was manipulatable via technological tinkering. It is this dual meaning of the prototype—the making and testing of alternatives and the normative modeling of a particular alternative—that I wish to highlight. Specifically, my concern is with the broad appeal of prototyping (the process) at the cost of attending to the prototype (the normative model that prototyping enacts). Allow me to make this point clear by returning once more to the phenomenon that was the impetus for this book; the vision of democratized innovation concentrated in the ideals and tools of making.

Making seemed malleable enough so that various ideals and political projects could be attached to it. But why was this the case? In this book, I have situated making historically to show its rise (and its supposed fall) in relation to a growing disillusionment with earlier promises and models of tech innovation—a disillusionment expressed by researchers, members of the tech industry, and media alike. What was so powerful about making in this very moment was the stipulation that the very tools and techniques of technological innovation (prototyping, tinkering, and experimentation) would be democratized—made available to "everyone"—via the provision of open toolkits. While making grew out of a specific sentiment of loss and uncertainty particularly pronounced in the American tech industry (see introduction to this book), it was portrayed as universally applicable (thus malleable)—its particular contestation of innovation was modeled as an archetype. Such positing of "new" models as somehow more authentic alternatives to the status quo is of course not neutral—and this political process deserves our close attention. As I have shown throughout this book, the very displacements of technological promise—the attachment of technological promise to "other" sites and bodies reframed as the latest "prototypes" of intervention—are also what funnels investments, attracts government funding, and legitimizes old and new exclusions. The "invention of Shenzhen" as an open toolkit of exuberant scale that I have documented in chapter 3 has contributed to the stipulation of the south of China as a new model to economic development and innovation—it has co-produced China as a *prototype nation*, the positing of China as the raw material for innovation that the CCP has strategically leveraged for its own purposes to center China as a model for infrastructural and economic development in the regions along its Belt and Road initiative (Africa, Asia, Europe, and the Middle East) and to expand its political and economic reach in Asia. In August 2019, amid expanding and enduring democracy protests in Hong Kong (neighboring Shenzhen), the *South China Morning Post* published an article entitled "Beijing unveils detailed reform plan to make Shenzhen model city for China and the world," with the subheading "Beijing wants Shenzhen to become a model of 'high-quality development, an example of law and order and civilization.'" Shenzhen, so the article explains, would become "a new special economic zone to carry out bolder reforms" with "international organizations and big companies encouraged to set up branches or headquarters in the city."

It is remarkable how familiar this "new" project of the SEZ is; not only because of its iterative designs in the shape of opportunity zones, smart zones, and innovation hubs, but also because of its self-reference—the "new" SEZ functions quite like the "old." It is a familiar technique of neoliberal governance that works by instilling competition between regions over investment and by producing feelings of transformation and change—crucial in moments of political, social, and economic upheaval. This technique is deployed by authoritarian regimes and democracies alike; it is currently deployed by the CCP to model Shenzhen as a high-tech, clean, civil, and future-oriented hub of innovation positioned in distinction from neighboring Hong Kong, which the CCP portrays as chaotic, violent, and irrational; and it is also deployed by the American Republican Party to associate *feelings* of opportunity and justice with its leadership.

As technology researchers, we must reckon with our own complicity in the violence legitimized by such displacements of technological promise when we attach feelings of hope, of countercultural resistance, and intervention to certain sites and bodies, but not others. We must challenge both the enduring neoliberal doctrine of inevitability *and* its refashioning as entrepreneurial life.

Contesting Inevitability

Is making just another story of capital's reach, another example of the subsumption of life by the machineries of capital? Indeed, disappointment over making's co-option by various structures of power surfaced repeatedly, even while making was still widely celebrated. Some asked whose empowerment it was. Maker- and hackerspaces seemed to advance the interests of masculine and technological elites rather than constituting a space for "everybody" as its prominent (and mostly male) figures claimed. A series of controversies contributed to a growing sense of betrayal and co-option. Most prominently, in 2013, the American 3D printing company MakerBot was sold for more than USD 600 million to Stratasys, and as a result, it closed significant parts of its open source repository.[12] About a year earlier, O'Reilly's Make (which would later become Maker Media) announced that it accepted several million USD in funding from the US military agency DARPA.[13] These developments seemed to draw a line between an older, purer, more subversive hacker movement and the newer version centered on making, which was thought of as co-opted by capitalist greed and corporate control.[14] Making became emblematic of how the juggernaut of capitalism can incorporate—even feed on—the critiques and social movements leveled against it.[15]

As the logics of finance capture even differential yearnings for alternative worlds, it is tempting to see capital as subsuming all intervention, as co-opting the maker movement and upending its ideal. But what would go unnoticed if we ended there, on disappointment, co-optation, the end of intervention and resistance? As a feminist ethnography, this book is committed to "noticing" what is otherwise ignored, or—as anthropologist Anna Tsing put it so well—to "keep sight of the continuing hegemony of scalability projects while immersing ourselves in the forms and tactics of precarity."[16] What has woven throughout this book are the knotted tangles of entrepreneurial life—and the gaps between its threads. I have traced displacements of technological promise across temporal and spatial scales, examining both what is and what could be otherwise. I also, in the words of historian Michelle Murphy, have often turned my "glance sideways," noticing that technology too can be otherwise. "There is no functionalist alignment between dreaming futures and financial logics," Murphy reminds us, "there are still many wells of unaligned dreaming that capital fails to register, and that, therefore, make other future imaginaries possible."[17]

What drew me more than a decade ago to sites of making was their stubborn—some might say naïve—insistence on working against the neoliberal ideology of inevitability. At times making traffics in idealism, telling seductive stories of success and techno-solutionism. But the insistence that *things can be otherwise* felt (and still feels) to me like an important intervention into the ideology of inevitability that pervades technology research and design itself.

The neoliberal doctrine poses economization—the expansion of capitalist production and capital's subsumption of life—as inevitable. As Margaret Thatcher notoriously said in the 1980s, "there is no alternative" to capitalism; at an Occupy rally in the 2010s, Slavoj Žižek posed that, "It's easy to imagine the end of the world—an asteroid destroying all of life, and so on—but we cannot imagine the end of capitalism."[18] Logics of inevitability have shaped technology research, design, and engineering fields for decades. Processes of economization within the global political economy seem like universal structures and meta-narratives; this "broader socioeconomic environment," within which technology use and design unfold, seems inevitable: "designers and technologists might not have a lot of leeway in reshaping" it.[19] And while some strands of design and computing research[20] have shown how technology design can be fundamental to (rather than distinct from) activism and political mobilization, broadly speaking, these fields continue to rely on problem-solving techniques and the ideals of techno-solutionism. There is a reluctance to address or even notice how the very methods of design and tech production—even critical ones—reproduce systemic issues of inequality and exclusion.[21] We co-produce the ideology of inevitability. I invoke, here, purposefully an empowered "we"— those of us who are designing, researching, or writing about technology. In these worlds, I have often encountered people who told me that technological progress was inevitable. In its most recent form, this narrative focuses on stories of data science, AI, autonomous vehicles, tech entrepreneurship, smart cities, and IoT. People say that universities and companies *have* to follow this trend (by retraining workers, updating curricula, changing research programs) in order to remain competitive, to attract customers or students or donors or investment. The stories we tell ourselves (and others)—about technological progress, competitive advantage, excellence, leadership—enable the ideology of inevitability by normalizing the belief that we *have* to follow suit.

This sense of inevitability within the worlds of technology research and design is deeply intertwined with an enduring belief in meritocracy, fueled by the heroic myths of the solitary technical genius. This is left over from early visions of information technology. After the Cold War, new narratives emerged: the rise of a networked information society, a knowledge economy, a creative class were seen not only as revolutionary but as *inevitable*—as part of a natural telos.[22] These narratives framed societal challenges as technological problems that could be solved by technological experts.[23] But the techno-fix, and the promise of economic development to which it is linked,[24] has masked continued colonial exploitation, resource extraction, racial discrimination, and economization of life.[25] In other words, the idealistic visions of an information society and of a global creative class and the democratized digital future they would bring have *re-legitimized* ideals of modern progress and hopeful forward movement precisely as these ideals were being contested; the promises of a technological utopia worked to shut down the movements toward decolonization that might have dismantled them.

Various versions of making reproduced some of these (often empty) technological promises. But if we consider making through its historical specificities and contingencies as this book has made it its central task (rather than taking the universalizing claims of some of its loudest advocates at face value), we notice the cracks that have been appearing for some time in the seemingly inevitable and endlessly churning wheels of capitalism and modernization. If we attend to the nurture of entrepreneurial life, we notice the vulnerability of modern progress. It took work to imbue making with a feeling of a shared capacity to contest inevitability; studying the production of this very feeling allows us to notice the differential yearnings for alternatives to the processes and structures of neoliberal capitalism.

Making provided scholars and entrepreneurs alike with a toolkit for feeling subversive; it provided a sense of agency in a seemingly inescapable neoliberal world, where alternatives—to social injustice, market capitalism, Western hegemony, labor precarization, and feminization of white-collar work—appeared out of reach. The fascination with making by critical scholars, governments, corporations, universities, and investors stems from this yearning for alternatives. These yearnings for alternatives fueled the promise of making that this book has examined. The scholar, the critic, the researcher—myself included—have been tangled up in the socialist pitch of entrepreneurial life. None of this is new, of course; we have long been complicit in recognizing only certain sites and places as "truly" countercultural. In order to contest inevitability, we must acknowledge our complicity in recognizing only certain practices, actors, and sites as authentic examples of resistance and intervention. There was nothing inherently countercultural about making or open source hardware. Imbuing technology with countercultural affect requires labor, difficult labor, non-glamorous labor, as I have shown it in chapters 5 and 6. This labor can be isolating; critiques of the status quo are often seen as "negative," inconvenient, and destructive.[26] Technological promise endures, because it is painful to notice our world in ruins. We cannot contest inevitability by replacing the utopia of neoliberalism with its seeming underdog—the countercultural hero who fuels the machineries of finance capitalism by creating affect and promise for technological interventions. And it is this questioning of the heroic notions that dominates discourse about technology that I have attempted to undertake in this book; namely, that the work of prototyping sought to create a seemingly authentic and renewed connection between man, machine, and economy, while hiding from view the sites and bodies that sustain this very promise of authentic connection. We can contest inevitability if we shift our "orientation" away from objects imbued with such promise. Once again I channel Sara Ahmed, who reminds us in "Orientations Matter": "to say that orientations matter affects how we think 'matter.' Orientations might shape how matter 'matters.'"[27]

What if we reoriented our commitments to justice? What if we deviated from the ideal types, from the prototypes of intervention that pitch so well?

What if we stopped following around objects of technological promise? What would it mean to prototype alternatives then?

When I traveled to Ghana in the spring of 2018 for a new research project (one that might occupy me for the next ten years), I met Ghanaians who felt both hope and despair about China's intensified investment in Africa. Colonialism endures in contemporary geopolitics, in how China construes its hegemonic reach (as distinct from the Western colonial "past"), in the promise that democratized innovation will bring justice for the oppressed, exploited, and formerly excluded, and in old and new orientalist tropes of othering. In Ghana, I also met Chinese who had come to Africa hoping to escape social injustice and inequality back in China; they too felt both hope and despair. They were living the dream of a newly empowered China, but at the same time, they were confronted with old and new racialized exclusions: "You know, we are not treated like the Westerners here," one of them told me. "They don't treat Chinese with the same respect. They treat us just like the white people [do]. To them I am low quality." Colonial pasts cast their shadows wide, and these shadows still orient our various yearnings for alternatives.

As I complete the edits to this book, China and its stature in the world are drastically different than they were when I began this research a decade ago. Today, the idea that China lags behind the West—the claim that China's early open source advocates sought to dispel—seems laughable. As Western media and governments around the globe agree, China is the new global economic superpower of the twenty-first century. And still, the Chinese Dream and China's investment efforts in Africa and other regions along the Silkroad 2.0 have not fueled a cultural imaginary like the American Dream; unlike midcentury America's promises of foreign aid, economic development, technological innovation, and democracy, China's foreign investment and thriving economy are not imbued with the same affect of modern, technological promise.

I have shown in this book how such denials to live the kind of future making and dreaming that is attached to white bodies and sites lead to further violence; such denials fuel the demand to self-sacrifice on behalf of the nation, the proliferation of precarious conditions of work and life, and the endurance of racism and sexism. But I have also shown that the endurance of colonial, racialized, and gendered othering are challenged on an ongoing basis, rather than passively endured. These challenges often go unnoticed for they do not pitch "well" as acts of countercultural heroism and outright resistance. And yet, it is exactly these slow, nagging, and ambivalent challenges to the displacement of technological promise that demonstrate a gap in the seemingly endless cycles of progress and ruin, promise and violence.

NOTES

Epigraph

1. "Prototype," *Oxford English Dictionary*, accessed September 2018, https://en .oxforddictionaries.com/definition/prototype.

Chapter 1. Introduction: The Promise of Making

1. Suchman, Trigg, and Blomberg (2002), 164.

2. Mellis and Buechley (2014).

3. Kaziunas, Lindtner et al. (2017); Kaziunas, Ackerman et al. (2017).

4. Anderson (2012).

5. As feminist and critical race scholar Sara Ahmed shows, happiness is socially constructed, connected to certain objects and life choices while being disassociated from others. Ahmed (2010a).

6. Berlant (2011).

7. As I will show, this shifting imagination of China's place in technology innovation, design, and engineering was co-produced by several processes: the CCP's experiments in national governance and global leadership (which included the strategic appropriation of China's colonial pasts), a growing disillusionment with existing approaches to design and tech innovation (a feeling that was particularly pronounced in the United States and that gained in intensity over the years following the global financial crisis 2007–2008), and shifting geopolitics and tensions in the global political economy that found their expression in debates over the ethics and politics of technological innovation.

8. These contemporary displacements of technological promise are historically constituted by economic development technologies and discourses that portrayed the "poor" and informal market practices as a key ingredient for future market success (Elyachar 2005). Such practices of what Karl Marx referred to as "primitive accumulation" have been—as anthropologist Julia Elyachar shows in ethnographic detail—integral to the functioning of neoliberal capitalist markets. While they produce violence and dispossession, these processes of economization are imposed through "more subtle processes" through which "the working poor . . . come to accept one version of the market as the only possible market" (for instance via "valorizing the cultural practices of the poor as a form of social capital" and "accumulation by debt"). See Elyachar (2005), 6 and 29.

9. Cross (2014).

10. Dirlik (2012b), 15. Prototype nation is different from (but lies adjacent to) Daniel Kreiss's use of the term "prototype politics" (Kreiss 2016). While Kreiss focuses on the uptake of digital media, data, machine learning, and analytics for contemporary political campaigning by the Democratic Party in the United States, my use of the term prototype is specifically concerned with the role promises and techniques of technological experimentation play in shifting geopolitics and modes of governance. Prototype nation discusses both the affect of technological experimentation *and* its appropriation for neoliberal governance.

11. Edward Said shows how the Orient was "reduced to nations without territory, *patrie*, laws, or security . . . waiting anxiously for the shelter of European occupation." See Said (1978), 179.

12. Avle, Lindtner, and Williams (2017); Irani (2018).

13. Saxenian (1996).

14. Florida (2002).

15. Parreñas (2018), 109.

16. Cross (2014); Easterling (2014); Murphy (2018); O'Donnell, Wong, and Bach (2017); Ong (2006); Parreñas (2018).

17. I am interested, in particular, how what Berlant (2011) calls "cruel optimism"—the holding on to the ideal of the good life despite its unattainability—is further enabled by digital technology. See also Benjamin's (2019) notion of "technological benevolence"—the portrayal of technologies through images of "do good," diversity, and progress.

18. Hecht (2002).

19. Anthropologist Aihwa Ong defines neoliberal exceptions as the selective adoption of neoliberal forms such as the creation of special economic zones and the imposition of market criteria on citizenship within an illiberal mode of governance. Ong (2006).

20. Stoler (2016), 5.

21. OCT (Overseas Chinese Town) Loft was established in 2001. Former factory spaces and warehouses were converted into chic lofts to attract tourists, families, flexible workers from entrepreneurs to designers and artists to work and play in its myriad of coffee houses, design studios, bookstores and small libraries, coworking spaces, and art galleries.

22. RMB (Renminbi) is the name of the Chinese currency.

23. The State Council (2015a and 2015b).

24. Xinhua News (2015).

25. The literal meaning of chaihuo (柴火) is "firewood" or "to stoke flames." The headline (李克强鼓励"创客"小伙伴：众人拾柴火焰高 Li Keqiang guli "chuangke" xiao huoban: zhongren shi chaihuo yangao) is a clever play on words and can also be translated as: Li Keqiang encourages young "makers": when everybody gathers fuel, the flames are higher (idiom; the more people, the more strength).

26. Ibid.

27. The State Council (2016).

28. Ibid.

29. See, for instance, Boltanski and Chiapello (2018).

30. Dougherty (2015), paragraph 2.

31. Ibid., paragraph 8.

32. The neoliberal ideology is synthesized by Margaret Thatcher's notorious phrase of the 1980s that "there is no alternative" to capitalism.

33. McRobbie (2016).

34. Roitman (2013).

35. The scholarship on this topic is rich and I can't possibly do justice to the wide range and nuance of topics and arguments here. Some of the work that has specifically informed my own thinking in this space: Chakrabarty (2007); Escobar (1994); Comaroff (1985); Ferguson (1994); Mamdani (1996); Parreñas (2018); Rabinow (1995); Stoler (1995); Tadiar (2009).

36. Feminization refers to the shift in gender and sex roles and the incorporation of women into a group or profession once dominated by men and the subsequent devaluing of work.

37. "Economization" refers to the rendering of "things, behaviors, and processes as economic," see Callon, Millo, and Muniesa (2007), 3. It is historically contingent and fundamental to the creation of markets, political leadership, and control; see Hecht (1998).

38. The conversion of life to capital—the framing of human existence in economic terms—has been traced historically along various axes. Some scholars situate its origins in the European enlightenment and modernization project (e.g., Foucault, 1980; Rose, 1996) and colonial

experiments (e.g., Ferguson, 1990; Mamdani, 1996); others see it as expanding during the Cold War and its aftermath (e.g., Anagnost, 1997; Murphy, 2018; Ong, 2006; Rofel, 2007). This kind of commodification—including self-commodification—is theorized as intensifying with the 1980s neoliberal techniques of governance, which call upon individuals to capitalize on their own capacities, to make themselves valuable to national economies and global finance capital. See Harvey (2007a); Brown (2015); Feher (2018).

39. This was a discursive shift away from the enduring techno-optimism that still pervaded the industry and public imaginary during the years following the dot-com crash in 2000, which—as Neff (2012) explores in great detail—was marked by a normalization of precarious work and risk in the tech industry; i.e., taking risks was portrayed as both "safe, natural, and routine" and even "cool," and as necessary for innovation (2–3). In the process, Neff shows, risk was privatized—corporate risk was loaded onto individuals, who were celebrated as innovative risk-takers and entrepreneurial adventurers. The ten years following the financial crisis in 2007–2008 witnessed a growing suspicion of the celebratory embrace of risk-taking and precarious labor as debates over digital labor exploitation, sexism, and racism in the tech industry, and surveillance technology began to pervade media and social media debates.

40. Tadiar (2013), 24.

41. Rao (2012).

42. Pacific Standard (2017).

43. Geier (2016).

44. Tolentino (2017).

45. Turner (2006).

46. Ibid., 209.

47. Kelty (2008), for instance, argues that software geeks came together through a recursive public, continuously made through the writing of code, i.e., the very social, economic, and political infrastructure of geek livelihoods. See also Coleman (2013).

48. This is not to say that there aren't numerous female-identified people writing, thinking, and working on maker and open source hardware–related projects. My concern, here, is with how their contributions were for the most part, with some exceptions (e.g., the recognition of prominent women such as Limor Fried, Naomi Wu, and Leah Buechley), not seen as "technological" or "innovative" or as "making," exactly because they did not follow a dominant view of what counts as technological expertise and creativity. Many of the women who have contributed centrally to technology production go unnoticed and unnamed, their voices and approaches often "heard" less than their male counterparts—a topic I will take up centrally in chapters 4 and 5. For excellent related work on gender in hacker and open technology production, see also Dunbar-Hester (2014); Eglash (2002); Fox, Ulgado, and Rosner (2015); Rosner, Shorey, Craft, and Remick (2018); SSL Nagbot (2016); for the reworking of computing history and women's central role in technology innovation, see, for instance, Hicks (2018); Rosner (2018); for gender in FLOSS comunities, see specifically Nafus (2012); Nafus, Leach, and Krieger (2006); Lin (2005); for the limits of peer production and the democratizing potential of open technology, see Dunbar-Hester (2020); Kreiss, Finn, and Turner (2011); Powell (2012).

49. Anderson (2012), 25–26.

50. The technique of pitching has traveled far and way beyond the world of start-ups, infiltrating various professions and work, even outside the tech industry. It has become common, for instance, for university professors and graduate students to be called upon to "pitch" their scholarship or teaching skills to potential donors, undergraduate student researchers, parents, and administrators. Likewise, demands are placed on employees and middle management in "traditional" industries (i.e., industries other than the tech industry) to reorient toward their work and company with an "innovator's mind-set," "entrepreneurial agility," or "digital skill set"—a topic I explore in greater depth in chapter 6.

51. Brown (2015); Feher (2018). While many began seeing the maker movement as a "failed" experiment, especially following *Maker Media*'s announcement that it had to shut down its operations due to financial hardship in 2019, it is precisely the socialist pitch that making helped sustain that endures until today. Making imbued the socialist pitch with renewed energy at a time when the long-held promises of digital tech production were being sharply critiqued. One of the key legacies of the maker movement is the establishment of an affective link between technological experimentation and finance capitalism.

52. I draw from a line of research that has challenged the belief in the rule of markets and shown through careful ethnographic and historical analysis what economization does and how it is reproduced and maintained; see, for instance: Hecht (1998), (2012); Collier (2011); Elyachar (2005); Mitchell (2002); von Schnitzler (2016); Murphy (2018); Zaloom (2010).

53. Here I align with scholarship that has challenged any reliance on "neoliberalism" as an overarching, explanatory frame. Instead of relying on analytical universals assumed to work the same everywhere, a body of scholarly work, largely rooted in long-term ethnographic engagements and careful historical analysis, has shown historically specific processes of economization rooted in complex interweaving of past and ongoing shifts in the political economy that unfolds through temporal and spatial scales; see, for instance, Hecht (1998), (2012); Collier (2011); Zhang and Ong (2008); Mitchell (2002); Murphy (2018); Neff (2012); Ong and Collier (2005); von Schnitzler (2016); Tadiar (2013); Tsing (2015); Zaloom (2010). This work demonstrates the arrangements and exclusions produced by various "mobile calculative techniques of governing" that "become translated, technologized, and operationalized in diverse, contemporary situations": Ong (2006), 13.

54. My specific focus is on how particular technological ideals of openness, inclusion, and participation tied to open source and peer production imbue processes of economization with affect. I build here on Michelle Murphy's theorizations of the economization of life as an active process, accomplished through "quantitative practices [of economization] that are enriched with affect"—what Murphy calls "phantasmagrams," i.e., "the felt and astral consequences of the social science quantification practices such as algorithms, equations, measures, plans; forecasts, models, simulations, and cascading correlations" that enrich economization with "affect, propagate imaginaries, lure feelings"; Murphy (2018), 24.

55. Callon (1998); Callon et al. (2007); Hecht (2012).

56. Hecht, building on Callon, elucidates that, despite claims to the contrary, "politics and economics remained tightly bound" . . . such that by "invoking the 'free market' imperial powers could continue to dominate former colonies" after independence, at a moment where their global domination was being contested: Hecht (2012), 35.

57. Scott calls this "state simplification"; Mitchell (2002) and Callon (1998) describe it as "enframing"; Dourish (2007) and Suchman (2006) analyze adjacent quantification and simplification measures and techniques in technology design and research, for instance via "interfacing" and usability engineering.

58. See, for instance, Terranova (2000); Fuchs (2014); Scholz (2012).

59. Irani (2019).

60. Sims (2017).

61. Benjamin (2019), 13.

62. Processes of economization are accompanied by a corresponding decline in collectivism, and increases in individualization and digital technologies are no exception. STS scholars document the processes of economization and corresponding depoliticization from the fragmentation and diffusion of politics into the mundane infrastructure of water meters (von Schnitzler, 2016) to the recasting of mineral resources as commodities (Hecht, 2012). Social agency is reframed as being accomplished not through political but economic action. Scholars have identified processes of depoliticization in many regions, from South Africa to postsocialist China; the shape of

depoliticization is contingent on specific historical, regional, and social contexts. It is the promise of economic agency that masks the endurance of colonial exploitation and racial discrimination; see, for instance: Hecht (2012); von Schnitzler (2016); Parreñas (2018); Stoler (2016); for China specifically, see: Anagnost (1997); Hui (2011); Ong (2006).

63. Information and STS scholar Christo Sims (2017) explores this in the context of educational reform whose framing as a "technocratic exercise" depoliticizes educational reform itself. Critical media scholar Wendy Chun shows how social media platforms advance depoliticization for they "do not produce an imagined and anonymous 'we' (they are not, to use Benedict Anderson's term, 'imagined communities') but rather, a relentlessly pointed yet empty, singular yet plural YOU . . . New media are N(YOU)media; new media are a function of YOU. New media relentlessly emphasize you . . . YOU is both singular and plural; in its plural mode, however, it still addresses individuals as individuals"; see Chun (2016), 3. Collective political action is fragmented via "micro-economic devices"; see Collier (2011).

64. Turner (2006).

65. Turner (2013).

66. Suchman (2011).

67. Hirosue, Kera, and Huang, (2015); Lin, Lindtner, and Wuschitz (2019); Lindtner and Lin (2017); Nguyen (2018); Sivek (2011); Turner (2018); for a contestation of American-centric discourse of making, see for instance, Hertz (2014); Chan (2014); Lindtner (2015).

68. Dourish and Mainwaring (2012).

69. Saxenian (1996).

70. Yang (2016a).

71. Neff (2012).

72. Tsing (2000), 53.

73. Cross (2014).

74. Ibid., 8.

75. Berlant (2011).

76. Rofel (2007), 12.

77. Sebastian Heilmann describes how Xi Jinping's ambitions to cement party leadership (and himself as the party's paramount leader and ultimate decision maker) express themselves in an attempt to reconnect the party to its historic roots (Leninist-style party discipline in particular) *and* a discourse that points toward the future. Heilmann describes this as a shift away from earlier experimental and "explorative" modes of governance toward "authoritarian innovations"; Heilmann (2016).

78. With technopolitics, the historian Gabrielle Hecht refers to the "strategic practice of designing or using technology to constitute, embody, and enact political goals": Hecht(1998), 15. I build on her definition to describe the strategic political usage of specific technologies and technological practices that were already imbued with a particular affect of promise and future making.

79. 中国梦 (zhongguo meng), earlier often translated as "China Dream," see, for instance, Callahan (2013). Communication scholar Aynne Kokas shows in great detail the role that transnational partnerships, at once symbiotic and competitive, between Hollywood and China's expanding media and cultural industries have played in the emergence of the Chinese Dream as a national branding strategy that utilizes an imaginary of creativity and futurity; Kokas (2017); communication scholar Fan Yang zooms in on the role creative and cultural industry policies have played in the CCP's strategies of branding the nation as a global producer of IPR-eligible brands and by extension a nation of dreamers capable of building and living the Chinese Dream, a vision that—as Yang argues—"conforms precisely to an 'American middle class' version of the 'good life,'" which Friedman has seemingly 'denied' China—a denial that numerous Chinese commentators have seen fit to contest by endorsing Xi's national dream in turn": Yang (2016a), 178.

80. Yang (2014b); see also Wielander and Hird (2018).

81. Peidong and Lijun (2018).

82. Callahan (2013), 8.

83. Yang (2014a), 47.

84. Ong (2006).

85. Jones (2011).

86. Ibid.

87. Dirlik and Meisner (1989).

88. Ibid., 18.

89. Ibid.

90. Anthropologist Lisa Rofel (1999) argues that the power of the communist party-state rested on its "ability to create and sustain a new mythos of polity, a new moral discourse of the nation . . . that framed a story of double loss and overcoming: the displacements of subalterns from their rightful place as inheritors of the fruits of their own labor and the loss of the nation to foreign powers": Rofel (1999), 26.

91. Dirlik (1989), 33.

92. Shenzhen was understood as a laboratory for national economic change, and its boundaries (as well as its success) helped appease senior members of the party, who ascribed firmly to a Mao doctrine that was highly suspicious of the economic opening and China's engagement with foreign entities. Cartier (2002); O'Donnell et al. (2017).

93. Gallagher (2005).

94. Chumley (2016); Florence (2017); Wilcox (2018).

95. Gallagher (2005), 14.

96. Chumley (2016).

97. Ibid.

98. Yang (2016a).

99. See, for instance: Dong (2019); Zhang (2019); Huang and Sharif (2017); for youth unemployment and perceptions of inequality in post-socialist China more broadly, see, for instance, Whyte (2012), (2010).

100. Parreñas (2018), 109.

101. See, for instance: Annany and Crawford (2016); boyd, Levy, Marwick (2014); Eubanks (2018); Gillespie (2014); O'Neil (2016); Metcalf, Moss, and boyd (2019); Pasquale (2015); Whittaker et al. (2018); Zuboff (2019).

102. It is even harder to notice various forms of violence when technology production is celebrated as democratized and open; exclusion and exploitation become a seeming non-issue, since individual empowerment is seen as a natural outcome of open technology. See also Benjamin (2019), Dunbar-Hester (2020).

103. Avle et al. (2017); Rosner (2018).

104. Dourish and Mainwaring (2012); Philip, Irani, and Dourish (2012); Irani et al. (2010); Taylor (2011).

105. Stoler (2016), 27.

106. By paying attention to the enduring influence of coloniality, Yang moves forward a project of decolonization, "which grounds itself in the material struggles of the Third World, from which China has emerged as a key figure since the 1955 Bandung Conference in Indonesia": Yang (2016a), 173–174.

107. Bach (2017).

108. Florence (2017).

109. Huang (2017).

110. O'Donnell (2017).

111. Bach (2017), 31.

112. O'Donnell (2017), 110.

113. Bach (2017), 31.

114. Neff (2012).

115. Ibid.

116. I deliberately use the male pronoun here to highlight the male dominance of start-up and computing cultures and their continuous enactments of masculinity, a point I return to in greater detail later; see also Dunbar-Hester (2014); Kleif and Faulkner (2003); Oldenziel (1991); Wajcman (1991).

117. Tsing (2000), 118.

118. I align here with the work of the anthropologist Sareeta Amrute, who challenges the notion of the "cognitive worker" (the creative class, the knowledge producer, the innovator, etc.) as a "universal, unmarked subject" and urges instead to consider how "race and class constitute a global terrain for cognitive work": Amrute (2016), 2–3. Racialized bodies, Amrute shows, provided "resources"—not unlike the happiness workers I describe in this book—for white Germans working in the tech industry to come to terms with the increasingly precarious conditions of their own jobs.

119. I align here with Marisa Cohn's research on the largely invisible labor that goes into maintaining scientific legacy (Cohn 2016) as well as Sareete Amrute's analysis of how highly skilled Indian tech workers in Germany endure racialized othering for the promise to build a new, global India (Amrute 2016).

120. Hecht (2002).

121. Stoler (2016).

122. Tsing (2004).

123. Drawing from Michel Foucault and those who have extended and critiqued his work, I pay attention to the recuperations, the reanimations, and the recombinations of past formations in the present. This approach differs from "tracing" the past's continuation in the present. Rather, it means the study of how multiple forms of power operate at once; see, e.g.: Foucault ([1966]1994); and as applied to considerations of race and colonialism: Stoler (2016). This approach challenges any politically charged and -motivated argument of China as the new global power (new empire), eager or destined to overtake the United States (old empire). The question to ask is thus not, will China take over the United States, but what are the specific recombinations of imperial formations in this present moment as the United States' and China's political leaders are engaged in a race over technological and global leadership? I also align with a body of scholarship that comes out of the interdisciplinary field of China studies and that has attended to ways in which China's colonial, revolutionary, and socialist pasts weave throughout China's recent and contemporary social, economic, and political processes—not in a linear fashion, but through recursions and reapproprations; see, for instance: Hoffman (2010); Rojas and Litzinger (2016); Merkel-Hess (2016); Yang (2017); Heilmann and Perry (2011).

124. Drawing from the historian Gabrielle Hecht, interscalar refers to the analytical movement between temporal and spatial scales: Hecht (2018).

125. See for instance: Boellstorff (2005); Faubion and Marcus (2009); Marcus (1995); Ong and Collier (2005); Zhan (2009).

126. Taylor (2011).

127. Hall (2016).

128. Chumley (2016); Justice (2012); Mullaney (2017); Ross (2007); Wallis (2013a); Wong (2013); Yang (2016a).

129. For instance: Ong and Collier (2005); Parreñas (2018); Tsing (2004).

130. Brown (2015); Feher (2018).

Chapter 2. Prototype Citizen: Colonial Durabilities in Technology Innovation

1. Wasserstrom (2009), 120.

2. The phrase "self-directed innovation" gained currency under Hu Jintao around 2006 and was picked up again in 2013 by Xi Jinping, who put central emphasis on building an indigenous innovation ecosystem.

3. Yang (2016a), 19.

4. Zizhu can be translated variously as self-governed, self-directed, and self-reliant.

5. Keane (2011), (2013); Flew and Cunningham (2010); Yang (2016a); Pang (2012).

6. Liu Shifa, fieldnotes by the author, 2004.

7. As with economic transformations since the 1980s, the party was centrally concerned with demonstrating that economic development and modernization would not mean Westernization or succumbing to Western hegemony as China became increasingly integrated into processes of globalization. To this end, the state introduced plans for cultural and creative industry development in order to position China as a newly modern nation—a producer of knowledge, ideas, and (by extension) of value.

8. I have previously explored this in Lindtner (2012); see also Keane (2011), (2013); Flew and Cunningham (2010); Yang (2016a); Pang (2012).

9. Turner elucidates how the techno-utopianist ideals that continue to be associated with digital technology today emerged out of a specific cultural context and a particular historical moment—at the intersection of the 1960s/70s American countercultural movements and the research labs that had grown out of the military industrial complex. From the early writings of the Whole Earth Catalogue in the late 1960s to the first publications of *Wired* magazine in the 1990s, an influential group of San Francisco Bay area journalists and entrepreneurs formulated the key ideological tenets that a generation of American tech producers followed: the belief that social change would come not from political mobilization but from experimentation with technology. This was a turn away from political action and toward the social and economic spheres as sites from which to launch social change; Turner (2013), (2006).

10. Lessig (2005).

11. See for analysis of the Shanghai Expo: Greenspan (2014); Hayden (2011); Hubbert (2017); Cull (2012).

12. XinDanWei's weekly "chit chats," informal talks given by members of XinDanWei's transnational network, were crucial in spreading ideas of open source and the free culture movement; connections made at these chit chats helped to initiate projects ranging from open source robotics and Arduino showcases to open design lectures and talks about sustainable urban development and design for good.

13. Jones (2011), 17. See also Russo (2009). Gina Russo shows how wenming was a foreign import "borrowed from the Japanese, who used the same characters in the late nineteenth century to define behavior that was specifically 'modern' and 'western,' thus maintaining the same connotations as 'civilized' in English. This notion of wenming was picked up by China at the beginning of the twentieth century with similar effect . . . the Nationalist government in the 1930s emphasized wenming behavior; it was often used in publications promoting the New Left Movement put forth by Chiang Kai-shek, a movement which encouraged people to be more hygienic and well-mannered in terms of clothes, food, behavior, and deportment": ibid., paragraph 1.

14. Anagnost (1997).

15. Jones (2011), 17.

16. Anagnost (1997); Chen et al. (2001); Chumley (2016); Yan (2008); Litzinger (1998); Rofel (2007); Schein (2000), (1997); Yang (2016a).

17. These experiments with inducing desires for self-upgrade among Chinese people were not a drastic rupture from the governance techniques with the Mao era, China scholars have shown. Rather, it displays a series of continuities with Maoist modernity, especially in the effort of the state to produce subjects that can most effectively strengthen the Chinese nation; see Wallis (2013a). Ann Anagnost shows the continuities with the 1920s, when ideas of the nationalist government centered around the idea that "to constitute a popular sovereign that is proper to the modern state, 'the people' have to be remade as part of a massive pedagogical project to rid them of the contamination of their 'feudal' past": Anagnost (1997), 11.

18. Rofel (2007).

19. Dirlik (2012a); Ong (2006).

20. Yan (2008), 132.

21. Chumley (2016).

22. This mode of governmentality is typically associated with finance capitalism; see for instance, Brown (2015); Feher (2018).

23. Rofel (2007), 12.

24. Yang (2016a), 24.

25. Wallis (2013a), 51; Ling (1999). It also positions the colonized as feminine and the West as masculine.

26. Anagnost (1997), 14.

27. Yang (2016a); as Yang shows, the CCP's attempt to denounce China's continuing historical practice of shanzhai production and substitute instead a Western-style culture of innovative production "worked to sustain rather than disrupt the unequal power relations [to the West], despite its ostensible intent to strengthen the nation vis-à-vis intensified global competition": Yang (2016a), 24. In other words, the copy was not shanzhai but the CCP's demand that its citizens produce "sameness" with the West by establishing a China-specific creative economy and innovation base mirroring that of the West.

28. Chumley (2016).

29. Ibid., 123.

30. Ibid., 125.

31. Lu and Perry (2015); Whyte and Parish (1984).

32. Zhang (2008), 26.

33. For a detailed study of the role of the hukou (household registration system) and the institutional legacy of socialist institutions in shaping migration and citizenship in the 1980s and '90s, see Solinger (1999); for the fashioning of minorities and migrants as entrepreneurial citizen subjects, see: Xiang (2016); Lee (2007); Oreglia (2014); Rofel (2016); Schein (1999); Wallis (2013b); for a timely analysis of how capitalist inequality shapes contemporary Chinese society, see Rojas and Litzinger (2016).

34. Wasserstrom (2009), 120.

35. Pseudonym.

36. Abbas (2000).

37. Lee (1999).

38. Leo Lee as cited in Abbas (2000), 775.

39. Cartier (2010).

40. Jones (2011). Callahan (2004); For a translation and detailed study, see Thompson (2011).

41. Jones (2011).

42. Callahan (2004).

43. Ibid., 570.

44. Translated from Chinese into English by the author.

45. The uptake of entrepreneurial life and open technology ideals aimed at achieving justice and parity with the West I describe here are not unlike the kinds of technological appropriations

Christian Sandvig observes in Indian reservations in the United States, which were aimed at achieving a sense of parity with the kind of infrastructural development more common in less remote corners of America. This "appropriation toward parity," Sandvig suggests, challenges the celebratory embrace of technology appropriation (common in user participation and open innovation research) as necessarily empowering, rendering invisible other aspirations that drive appropriation. Sandvig (2012).

46. Lindtner and Szablewicz (2011).

47. Since the economic reforms in the 1980s, the CCP has been centrally concerned with the question of how to retain the Chinese state's legitimacy as it transitioned from socialist to capitalist market principles. It sought to create continuity between capitalism and China's traditions and values by deploying a narrative, drawn from Confucianist philosophy, of China's constant development toward a harmonious society; see: Greenhalgh (2010); Ong (1999), (2006); Rofel (2007); in the context of the Chinese Internet: Lindtner and Szablewicz (2011); Harold and Marolt (2011); Qiu (2009). The government interventions aimed at controlling the Internet in order to limit Chinese netizens' potential political power ranged from its initiation of the "great firewall of China," which blocked access to certain websites and censored particular content, to the installation of control mechanisms on public computer terminals that limited certain types of searches.

48. Creative Commons (CC) is a non-profit organization headquartered in Mountain View, California, devoted to expanding the range of creative works available for others to build upon legally and to share. The organization has released several copyright licenses known as Creative Commons licenses free of charge to the public. These licenses allow creators to communicate which rights they reserve for themselves and which rights they waive for the benefit of recipients or other creators.

49. Lessig (1999), 7.

50. Coleman (2013), 198.

51. Mao, "Sharism."

52. Mao (2008).

53. Lessig (2008).

54. Ibid.

55. Lessig in *Freesouls*.

56. Baym (2018).

57. Brown (2015); Feher (2018).

58. Ahmed (2010a); Cooper and Waldby (2014); Cowen (2014); Ferguson (1990), (2006); Levien (2018); Sunder Rajan (2012); Tadiar (2009); Sanval (2007); Vora (2015); Roy (2010).

59. Berlant (2011).

60. Ahmed (2010a).

61. Justin Wang, Opening Speech, Beijing Maker Carnival, 2012, transcription by the author.

62. Ahmed (2010a), 125.

63. Ibid.

64. Anagnost (1997); Dirlik (2012b); Rofel (1999), (2007); Wallis (2013a); Yang (2016a).

65. Turner (2006), 245.

66. Coleman (2013); Douglas (2002); Ito (2009); Kelty (2008); Levy (2010).

67. Florida (2002).

68. For China-specific context, see for instance: Keane (2016), (2013), (2011); Yang (2016); in a translocal context: Avle et al. (2017); Avle and Lindtner (2016); Irani (2018), (2019).

69. McRobbie (2016), 15; see also Neff (2012); Martin (1994); Baym (2015); Gregg (2011).

70. My arguments here depart from Coleman's observations of free and open source hacking as a form of resistance to contemporary forms of neoliberal capitalism by intervening in the

mechanisms of IP and corporate monopoly, see Coleman (2013). While technologies certainly can play a role (and should play a role) in countering neoliberal structures of exploitation, we must attend to how the affect attached to technological resistance ironically precludes collective political mobilization by channeling political desires into economic productions.

Chapter 3. Inventing Shenzhen: How the Copy Became the Prototype, or: How China Out-Wested the West and Saved Modernity

1. Ito (2014a).

2. Ibid.

3. Ito (2014b).

4. You see this concern reflected in a course co-taught by Jonathan Zittrain and Joi Ito at the MIT Media Lab on "The Ethics and Governance of Artificial Intelligence" as well as in Ito's 2019 *Wired Ideas* essay, "Supposedly 'Fair' Algorithms Can Perpetuate Discrimination."

5. Davis (2016a); McRobbie (2016); Neff (2012).

6. For critical perspectives on the ethics of computing, and AI and data science in particular, that began emerging right around that time, see, for instance: boyd et al. (2014); Eubanks (2018); Gillespie (2014); O'Neil (2016); Metcalf et al. (2019); Pasquale (2015); Zuboff (2019).

7. Said (1978).

8. Berlant (2011), Cross (2014).

9. Yang (2016b), 382, 387. Narratives of fiscal orientalism, Yang shows, racialize the effect of neoliberal financialization by fusing the time of the Other and the time of debt; "China cannot exist in the present, but a Chinese future that promises to 'enslave' America . . . presents an unmistakable risk for the United States" (pp. 381–382). Racialized fears are channeled, in other words, to create both anti-Chinese sentiments and American national identity in a moment of contestation between nation-state and finance capital.

10. Morley and Robins (1995) describe how, during and following Japan's economic and technological advancement, it was depicted in American political discourse and cultural production (especially the genre of cyberpunk) as an oriental other more powerful than the American self, a "figure of empty and dehumanized technological power" (p. 170) placed into the future "so as to delegitimize its present status as a leading nation of technological advancement" (Yang, 2016b, 378). This form of techno-orientalism is discussed, for instance, in the following works: Chun (2006); Nakamura (2002); Ueno (1999); Roh, Huang, and Niu (2015). My aim is to contribute to a recent line of scholarship that calls for close attention to the shifts in orientalist othering in the Western, and particularly American, imagination amidst a growing realization of China-US interdependency since the global financial crisis: Tan (2015); Yang (2016b).

11. We should not understand the "Third World" as passive actor in this process, but attend to processes of co-production, often highly uneven and ambivalent. Escobar (1994), 7–8; see also: Ahmed (2010); Chakrabarty (2007); Elyachar (2005); Mitchell (1991); Mudimbe (1988); Said (1978); Spivak (1999).

12. Escobar (1994), 26.

13. I invoke here Mudimbe (1988), to highlight the contestations over a China-centric approach to innovation, and to attend to the endurance of colonial and orientalist tropes of othering that pervaded the discursive construct of a newly innovative Shenzhen that shaped the region and its people, while it never fully defined them.

14. Brown (2015).

15. While orientalism originated in Europe and was central to the production of European modernity (by distinguishing it from the "not yet" modern other), it was all but one-directional

with Asians as passive actors rendered as orientalists; for a detailed account, see: Escobar (1994); Dirlik (1997)—as Dirlik summarizes it: "Asians participated by ways of self-orientalizing, which in turn became an integral part of the history of orientalism . . . Euroamerican images of Asia became incorporated into the self-image of Asians in the process and in the end became inseparable from the impact of 'Western' ideas per se": ibid., 15. Roh, Huang, and Niu (2015) in a similar vein show that "techno-orientalism"—especially in its incarnations in the wake of neoliberal trade policies—is bidirectional, "its discourse mutually constituted by the flow of trade and capital across hemispheres," with Asian nations finding "in techno-orientalism an expressive vehicle for their aspirations and fears" (p. 3).

16. McRobbie (2016).

17. Davis (2016a).

18. Florida (2002).

19. Following the 2007–2008 financial crisis, massive government bailouts, support for faltering companies, emergency tax cuts transformed the burden from private debt into high levels of public debt; see Srnicek (2016).

20. Tsing, (2015), 20.

21. Berlant (2011).

22. Fuchs (2014); McRobbie (2016); Scholz (2012).

23. See popular critiques of design thinking, e.g.: Iskandar (2018); Jen (2018); Kolko (2017); Nussbaum (2011); Vinsel (2017); for scholarly debates on the commodification of Participatory Design: Bannon, Bardzell, and Bodker (2019); Bannon and Ehn (2012); Beck (2002); Kyng (2010); Sengers et al. (2005).

24. Escobar (2017).

25. Fisher (2020).

26. Pasquale (2015); Sandvig (2019); Escobar (2017); Zuboff (2019).

27. Jackson (2014).

28. Ibid.

29. Chan (2014).

30. *The Economist*, "The Redistribution of Hope"; cited in Greenspan (2014), xxv–xxvi.

31. Chan (2014).

32. Harvey writes that "postmodern thought did away with the possibility of better or alternate futures"—the loss of the alienated subject, replaced by the fractured self, also precluded the conscious construction of alternative social futures, Harvey argues. Harvey (1990), 54.

33. Greenspan (2014), xxvi.

34. The project of European modernization itself was one of turning to other regions in order to build the utopia of the First World.

35. Öniş (1991).

36. Johnson (1982), 9.

37. Ibid., 19.

38. Beeson (2009); O'Donnell et al. (2017).

39. Easterling (2014); Rossiter (2016); Yang (2016a); O'Donnell et al. (2017).

40. The Marxist scholar David Harvey argues that this broad rejection of utopianism has risen over the past two decades and stems from the collapse of specific utopian forms, both East and West. Communism has been broadly discredited as a utopian project and neoliberalism is increasingly seen as a utopian project that cannot succeed. Insofar as the geopolitical strategy of the United States can be understood as a form of spatiotemporal utopianism, it, too, is less and less convincing. Nevertheless, Harvey argues, "utopian dreams in any case never entirely fade away. They are omnipresent as the hidden signifiers of our desires." Harvey (2000), 195.

41. Bardzell (2018); Harvey (2000); Pal (2017).

42. We see this call to update our approaches to and understandings of resistance variously expressed in studies of speed, acceleration, supply chains, special economic zones, digital precarity, and tech labor; e.g., Easterling (2014); Harvey (2000); Dyer-Witheford (1999); Rossiter (2016).

43. Ong (2006), 98.

44. Easterling (2014), 60.

45. Gallagher (2005).

46. Jamie Cross, for instance, documents the uptake of SEZs in India since the 1990s; Cross (2014).

47. Easterling (2014); O'Donnell et al. (2017).

48. Easterling (2014), 25.

49. Ibid.

50. Ong (2006), 98.

51. Easterling (2014).

52. Bach (2017), 36.

53. Ibid.

54. Harvey (2007b).

55. Cross (2014), 9.

56. Ibid.

57. Easterling (2014).

58. Wallis and Qiu (2009).

59. Ho (2010), 1.

60. Zhang and Fung (2010).

61. Wallis and Qiu (2009).

62. Jeffery (2014).

63. Rossiter (2016), xiv.

64. Wallis and Qiu (2009).

65. Emphasis mine.

66. Cartier (2010).

67. Anagnost (1997); Ong (2006); Rofel (2007).

68. Greenhalgh (2010). Popular phrases like 西学为用，中学为体 (xixue wei yong, zhongxue wei ti, western learning for function, Chinese learning for substance), which emerged at the end of the Qing dynasty, in the mid-nineteenth century, drive home the idea that capitalism is ultimately intended to increase the power of the Chinese nation-state and that modernization was a site to renew Chinese culture; see: De Bary and Lufrano (2000); Greenhalgh (2010); Ong (2006). These phrases were used to reinforce the idea that capitalism would ultimately increase the power of the Chinese nation-state and to assert that modernization would renew Chinese culture; see also: O'Connor (2012); MacFarquhar and Schoenhals (2006); Jones (2011); Anagnost (1997); Dirlik (2012b); Rofel (1999); Wallis (2013a).

69. Dirlik (2012b), 120.

70. Dale Dougherty, Keynote at the 2014 Shenzhen Maker Faire, fieldnotes by the author.

71. Neff (2012).

72. Huang (2013a, 2013b).

73. Huang (2013a).

74. Ibid.

75. Huang (2013a).

76. Ibid.

77. Lessig (2005), 77, as cited in Philip (2005), 210.

78. Philip (2005).

79. Ibid., 200–201.

80. Hayden (2010); Philip (2005).

81. Simonite (2016).

82. Huang (2013b).

83. Ibid.

84. IDEO Patterns (2009–2011).

85. Pang (2012).

86. Booz & Company (2009).

87. Christenson (1997).

88. Cormier (2015).

89. See, for instance, the Design Justice Zine series produced by the collective formed around designjustice.org and Constanza-Chock (2020) for an excellent synthesis of many years of work dedicated to the topic.

90. The phrase is from a piece of musical propaganda by the same name, sung by Xu Qianya, which Mary Ann O'Donnell documents has been used by the municipal government since 2012, when it decided to further loosen the city's hukou requirements, with the campaign aiming to "graft a sense of place onto deterritorialized bodies": O'Donnell (2012).

91. Ole Bouman, Public Talk at the *2016 Shenzhen Maker Faire*, Shenzhen, China.

92. Ibid.

93. The website defines an "official board" as an "Arduino board that is directly supported by the official Arduino IDE, follows the Arduino layout we have standardized, is properly documented on our website, is properly licensed to bear the Arduino name and logo, is made by authorized manufacturers." In contrast, "Clones" are defined as "exact (or almost exact) replicas of Arduino boards with a different branding, i.e. they are usually named with some variation of Ardu- something or something–duino." See Banzi (2013).

94. Turner (2013), 82.

95. Excerpt of Massimo Banzi's talk at the 2013 Hardware Innovation Workshop in San Mateo, fieldnotes by the author.

96. See Benchoff (2016); Williams (2015).

97. Steffan (2015).

98. La Repubblica (2015).

99. Hayden (2010), 87.

100. Ibid.

101. Liang (2014), 168.

102. Shirky (2015), 51.

103. Wong (2013), 41.

104. Rosner (2018); Turner (2006).

105. Hayden (2010).

106. Ibid., Dunbar-Hester (2014);

107. Fields and methods of design and computing—from human-computer interaction to design thinking—are built on the promise that digital technologies would bring about the good life and a better future (an attitude that still persists in contemporary technology discourse and practice). The "digital" had been envisioned to revive the promise of modern progress in the moment as technological and scientific progress was seen with growing suspicion in the 1970s amidst the rise of a postmodernist aesthetic in the arts and the flowering of poststructuralist, feminist, and postcolonial critiques in the academy (see Harvey 1990). By the time shanzhai garnered the attention of designers and engineers, design and computing had begun to increasingly reckon with the poststructuralist, feminist, and postcolonial critiques of progress and development. While in the late 2000s, digital technology (and social media in particular) were still seen as sources of political agency and intervention (said to have fueled the Arab Spring and constituting an arsenal

of democracy), this very promise soon appeared absurd. By 2016, it became even more clear that Facebook had played a central role in manipulating US voters and that Twitter had given the newly elected US president a platform on which to display the erratic behavior typically associated with authoritarian leadership rather than with democracy and the free world. The attention to shanzhai thus coincided with this proliferating feeling that the tech industry was in crisis.

108. How much was at stake in this recuperation of the Media Lab via Shenzhen as an institute that "stood for" ethical hacking has become even more clear with the recent 2019 revelations of Media Lab's involvement and acceptance of millions of US dollars in funding from the financier and convicted sex offender Jeffrey Epstein (see my reflections on this topic in the conclusion of this book).

Chapter 4. Incubating Human Capital: Market Devices of Finance Capitalism

1. I use pseudonyms for names throughout this chapter to protect the identities of the people I worked with.

2. Sacks and Thiel (1998).

3. Hall (2016).

4. Thiel and Masters (2014).

5. Anderson (2012), 8.

6. Foucault (2008); Brown (2015); Murphy (2018); Feher (2018); Rose (1998).

7. Sims (2017), 3.

8. Feher (2018).

9. Brown (2015), 34.

10. Murphy (2018), 116.

11. Feher (2018).

12. Feher (2009).

13. Anderson (2012), 26.

14. Pettis (2015).

15. Neff (2012).

16. Ibid., 14.

17. Curtis and Sanson (2016); Fuchs (2014); Hall (2016); McRobbie (2016); Scholz (2012); Srnicek (2016).

18. Geron (2002).

19. Sims (2017).

20. Hall (2016).

21. Ibid., 20.

22. Ibid.

23. Murphy (2018).

24. Ibid., 115.

25. Sims (2017).

26. I refer, here, to a number of recent experiments with scholarly work and instution-building at the fringes or outside of the university (prominent examples include, for instance; Data & Society and the AI Now Institute). Despite ties to the tech industry, capital investment, and/or wealthy donors and foundations, these technology research institutes are understood as enabling scholars to be in touch with pressing societal and political matters and to make their research newly relevant. Portrayed as an alternative to the "slow" and "stodgy" world of the university, the promise of intervention and impact masks the demand to accelerate scholarship and the normalization of precarious conditions of scholarly labor that these institutions require for their own survival.

While they provide important interdisciplinary links between scholarship, policy, and industry as well as temporary employment, especially for young and highly precarious scholars, they help legitimize—alongside the neoliberal university itself—a move away from tenure-track employment, an increase in adjunct and short-term contract labor that in turn makes critical scholarship ever harder to sustain (as reseachers fear retaliation from the institutions that employ them and that serve first and foremost "student needs," corporate interests, market demands, and donor preferences); see Hall (2016).

27. Ensmenger (2003).

28. Ibid.

29. Ibid., 154.

30. Davis (2016a).

31. Callon (1998).

32. Ibid.; Hecht (2012).

33. Callon (1998); MacKenzie (2008); Mitchell (2002).

34. Feher (2018), 18.

35. Feher (2018).

36. Bannon, Bardzell, and Bødker (2019); Huppatz (2015); Hughes (2005).

37. Ames (2020); Sims (2017).

38. Feher (2018).

39. Many male hackers and computer programmers experience a feeling of pleasure and control, because past socialization has taught them to find comfort in technology; see for instance: Edwards (1996); Turkle (2005); Bray (1997); Kleif and Faulkner (2003); Wajcman (1991); Ensmenger (2003); Rankin (2018). Research on enactments of masculinity in China include, for instance: Hird (2009); Hird and Song (2019); Shen (2008); Song and Hird (2014).

40. This includes, for instance, widely taught and translated work such as: Brown (2009); Norman (2013).

41. von Hippel (2005).

42. Stories of crowdfunded products whose shipments are delayed, often by months, and of their founders, who often barely break even or lose money, abound. One of the most prominent examples is the hardware company Pebble Smart Watch, which raised a record-setting Kickstarter campaign (more than USD 10 million) and was long celebrated as one of the key success stories of the maker movement. This was until stories of late shipments, disappointed customers, and bad PR broke. A recent flurry of documentary films critically investigate the Silicon Valley mantra of "fake it, till you make it" by uncovering how millions and even billions of US dollars were invested in the fictitious stories of charismatic inventors; see for instance the HBO documentary *The Inventor* and the Hulu channel's original documentary *Fyre Fraud*.

43. Name of the program is anonymized.

44. Söderberg and Delfanti (2015).

45. Feher (2018), 21.

46. Murphy (2018), 7.

47. Wajcman (2016), 14.

48. Ibid., 17.

49. O'Donnell (2017), 41.

50. Ibid., 54.

51. Florence (2017).

52. Ibid.

53. Ibid., 86, 88.

54. Ibid., 98.

55. See Lorde (2007).

Chapter 5. Seeing Like a Peer: Happiness Labor and the Microworld of Innovation

1. I use pseudonyms for names throughout this chapter to protect the identities of the people I worked with.

2. Hochschild (1983).

3. Neff (2012).

4. Brown (2015); See also, Foucault (2008).

5. Brown (2015), 34.

6. To reiterate and as previously discussed, "economization" refers to the rendering of "things, behaviors, and processes as economic" (Callon, Millon, and Muniesa 2007, 3) and self-economization is the neoliberal demand that one convert the self into human capital, investing in various aspects of one's own life in order to make the self attractive to the machineries of finance speculation and investment; see Brown (2015); Feher (2018).

7. Eglash (2002); Kreiss, Finn, and Turner (2011); Dunbar-Hester (2014).

8. With the phrase *seeing like a peer* I build on James Scott's theorization of the processes and instruments of quantification, simplification, and measurement (from maps to surveys) that functioned as tools of governance for the modern European nation-state. I argue that the tools of governance widespread in finance capitalism tend to be less focused on simplification—as Scott found—and more on producing an affect of change and anticipatory promise; see Scott (1998).

9. On precarious conditions of work in the tech, media, and creative industries, see: Baym (2018); Curtis and Sanson (2016); Duffy (2017); Gray and Suri (2019); Gregg (2011); Lindtner (2017); McRobbie (2016), (2010), (2011); Neff (2012); Ross (2004); on questions of race, outsourcing, and transnational labor migration in the tech industry, see specifically: Amrute (2016); Nadeem (2013); Nakamura (2002), (2007); for definitions and theorizations of precarity see: Butler (2009); Cole and Hattam (2017); Lorey (2015), (2010); Neilson and Rossiter (2008); on tech labor and crowdwork see: Dyer-Witheford (1999); Irani (2013), (2015); Scholz (2012), (2016); Terranova (2000); Silberman, Irani, and Ross (2010); Qiu (2018); Raval and Dourish (2016); Wallis (2013b); on tech labor and logistics see for instance: Cowen (2014); Rossiter (2016); for critical analysis of "future of work" rhetoric in politics, labor, and technology activism see: Forlano and Halpern (2015).

10. Lazzarato (1996).

11. Tronti (1966). The concept of the social factory, formulated in the 1960s by Mario Tronti, extended to society as a whole Marx's notion of how capital "subsumes"—how it drives the "ever far reaching incorporation of sites and practices into processes of capital generation." As Tronti wrote in the 1960s, "at the highest level of capitalist development, social relations become moments of the relations of production, and the whole society becomes an articulation of production. In short, all of society lives as a function of the factory, and the factory extends its exclusive domination over all of society"; Tronti, *Operai e Capitale*, quoted in: Cleaver (1992), 137. The autonomous Marxists "moved away from the traditional Marxist focus on the immediate point of production, i.e., the factory, and included also distribution and consumption": Dyer-Witheford (1999), 67. By subsuming under itself not only production but also distribution and consumption, capital was understood as transforming all of society into a factory. By focusing on the critical but unacknowledged role played by the reproduction of labor in Marxist theory, autonomous Marxists performed a feminist critique of the political economy.

12. See for instance: Dyer-Witheford (1999); Fuchs (2014); Terranova (2000); Scholz (2012).

13. Florida (2002).

14. Castells (2009).

15. For an excellent critical analysis of various visions of the information society, see Webster (2006).

16. Bannon and Ehn (2012); Irani and Silberman (2013).

17. Zuboff (2019).

18. Srnicek (2016).

19. Eubanks (2018).

20. Berlant (2011).

21. Latour (2004); Anderson (2008). In a blog post, critical computing scholar Jeffrey Bardzell summarizes the sentiment that critique is "negative" in HCI scholarship: Bardzell (2018).

22. McRobbie (2016).

23. Ross (2014).

24. Brown (2015); Feher (2018).

25. Fuchs (2014), 227.

26. Neff (2012), 136.

27. Fuchs (2014), 228.

28. Nakamura (2011), 7–8.

29. Anagnost (1989); Hanser (2008); Litzinger (2016); Rofel (2016); Schein (1997); Solinger (1999); Wallis (2013a).

30. Rofel (2016), 171.

31. See, for instance, Fuchs (2014); Harvey (2007a).

32. Srnicek (2016).

33. To note but a few: Dong (2019); Gallagher (2005); Kuruvilla, Lee, and Gallagher (2011); Wallis (2013a); Qiu (2018); Rofel (2016); Lee (2007); Lee (1998); Litzinger (2016); Zhang (2019).

34. Rojas and Litzinger (2016).

35. Ahmed (2010a), 32.

36. Berlant (2011).

37. In contrast to myriad networking opportunities for start-ups and makers, there is little opportunity (and certainly no investment made) for happiness workers to connect and share experiences. Many of the women I met worked in relative isolation for the respective organizations that employed them. It was challenging to build solidarity with women in start-up teams, who were themselves fighting for legitimacy in a highly masculinized world of innovation.

38. Turkle (2005).

39. Edwards (1996), 171.

40. Ibid., 171–172.

41. The notion of the microworld is adjacent to Gary Allan Fine notion's of idioculture, i.e., cultural items in a small group that are central to the group's continuous existence: Fine (1979). My concern here is with how the microworld or idioculture of the incubator was produced by the displacements of technological promise (the recursive moves of technological ideals and hopes for interventionist change).

42. Dourish and Bell (2018), 773.

43. Ibid.

44. Ibid.

45. Dunbar-Hester (2014), 66.

46. Eglash (2002).

47. Wajcman (1991).

48. Ibid., 32–34.

49. See for instance: Edwards (1996); Dunbar-Hester (2014); Turkle (2005); Bray (1997); Hicks (2018); Kleif and Faulkner (2003); Wajcman (1991); Ensmenger (2003); Rankin (2018); Traweek (1988).

50. Wajcman (1991), 37.

51. Baym (2018); see also Cockayne's notion of "entrepreneurial affect" (2015).

52. Hochschild (1983).

53. Modrak (2015); Butler (1999).

54. Tsing (2015).

55. Scott (1987), 34.

56. Anna Watkins Fisher (2020), 23.

57. Ibid.

58. Ibid.

59. Murphy (2017), 134.

60. Bardzell (2018).

Chapter 6. China's Entrepreneurial Factory: The Violence of Happiness

1. I use pseudonyms for names throughout this chapter to protect the identities of the people I worked with.

2. Kücklich (2005); Fuchs (2014).

3. Forms of violence such as racialized and colonial othering as well as precarious conditions of work are endured because of the promise that self-upgrade into an entrepreneurial, worldly innovator would eventually lead to justice and the good life.

4. Ahmed (2010a), 8–9.

5. Ibid., 11.

6. Ibid., 29.

7. Chen (2017).

8. Wallis and Qiu (2009); Lindtner, Greenspan, and Li (2015).

9. Even though large segments of these companies were privatized in the 1990s, they still retain a small percentage of government ownership.

10. I am drawing centrally here on the detailed analysis of the past and continuous role SOEs played in Shenzhen's experimental development by Ting Chen in her 2017 publication *A State Beyond the State. Shenzhen and the Transformation of Urban China.* Chen challenges the more pervasive assumption that Shenzhen was planned by one central government entity. Instead, she shows, the Special Economic Zone was planned, developed, and built, through a multi-level urban planning and development approach, with each SOE experimenting with different approaches to urban form and living within their zones.

11. Ibid., 34. China Merchants is a Hong Kong–based shipping company affiliated with China's Ministry of Transportation. In 1979, it was charged by the state council to build an independent comprehensive zone for the port, industry, and trade—Shekou. Shekou then had its own departments for police, planning, and land administration, public health, education, and border control.

12. This included, for instance, the low-rise industrial buildings of Huaqiangbei (and elsewhere) built during the 1980s.

13. Chen (2017), 93–94.

14. Ibid., 29.

15. Maker-related events from workshops in local high schools and libraries to large-scale Maker Faires or similar show-and-tell events frequently attracted large numbers of local residents and especially middle-class families who sought after-school and weekend activities for their children that they associated with creative learning. Many of the families and parents I met over the years told me that maker-type activities were just what they wanted for their children—an educational pastime that taught children basic skills in science, technology, and math and that promised to pull them away from activities such as online gaming or passive Internet use, which many understood to be unhealthy for their children.

16. Stone (2014).

17. Chen (2017); Gallagher (2005); O'Donnell, Wong, and Bach (2017).

18. Schein (1997), 73.

19. Ibid., 74.

20. I use pseudonyms for names throughout this chapter to protect the identities of the people I worked with.

21. For purposes of anonymity and protection, I only use his self-chosen English name.

22. For purposes of anonymity and protection, I only use her self-chosen English name.

23. RMB (Renminbi) is the name of the Chinese currency; at the time of my research, RMB 3600 roughly converted into USD 560 and RMB 9600 into USD 1500.

24. See the CCTV series on the topic aired in October 2017: http://tv.cntv.cn/video/VSET100232480132/a655bc32eb3f4e61a2518a943c2b43df.

25. Brown (2015).

26. Simon (2018).

27. Kuruvilla, Lee, and Gallagher (2011).

28. Fuchs (2014); Scholz (2012); Terranova (2000).

29. See for instance, Irani (2013), (2015); Irani and Silberman (2013); Raval and Dourish (2016).

30. The successful attraction of business can be lucrative for provincial and municipal real estate and government entities as it often promises additional funding and/or support from the central government.

31. Hui (2010); Heilmann and Perry (2011); Yang (2017); Wasserstrom and Perry (2019).

Chapter 7. Conclusion: The Nurture of Entrepreneurial Life

1. IRS, "Opportunity Zones Frequently Asked Questions."

2. Aticelli (2019).

3. Ibid.

4. Neff (2012).

5. Sims explores this in the context of philanthropy, Sims (2017).

6. It is incredibly difficult to challenge such projects that employ a version of the socialist pitch I have described in this book; the appeal to social justice, fairness, and morality legitimizes projects of investment; see also Sims (2017). These projects persist despite their critiques in media outlets, despite their politicization—it is the aesthetics of morality and of public debate that enable these projects—they create a temporary buffer, what Sims (2017) calls "politicized buffer zones," that absorbs critique but leaves the promise itself intact. See also Irani (2019).

7. The website that expressed this support was taken down soon after Ronan Farrow's *New Yorker* article was published.

8. Philip (2005).

9. Right around the time Ito traveled to Shenzhen, he had updated the Media Lab's famous "demo or die" slogan to "deploy or die"—prototypes of intervention that promised real-world change and demanded of students to engage with the reality of industrial production and manufacturing.

10. The critique of critique ignores a whole body of feminist and critical race scholarship that has provided the analytical and methodological repertoire to challenge a position from nowhere and to move beyond distant critique; see, for instance, Ahmed (2017); Bardzell (2018); Benjamin (2019); Fisher (2020); Gibson-Graham (2006); Harney and Moten (2013); Haraway (1991); hooks (1990); Puig de la Bellacasa (2011); Rosner (2018); Joseph (2018); Stengers and Despret (2014); Tsing (2015), (2004). As this scholarship shows, the critique of critique comes from a distant

observer, positioned above and removed from those he renders as dupes: those who make a fuss, those who argue for critique as salutary, those who *need* critique—the women and men of color, the people viewed as less modern, the populations rendered disposable. Their mode of critique has always been embedded; it was never outside to begin with. With this book, I have aspired to join them in making a fuss, in committing to embedded critique, to acknowledging my anger, my own yearnings for other, better worlds—a tangle that pulls us not away from but deeper into the machineries of late neoliberal capitalism.

11. Escobar (2017), 6.

12. See media coverage on the topic, for instance: Clay (2013); Etherington (2013).

13. O'Reilly Press Release (2012).

14. Maxigas (2012).

15. Boltanski and Chiapello (2018); Söderberg and Delfanti (2015).

16. Tsing (2015), 42.

17. Murphy (2018), 134.

18. The same quote, "it is easier to imagine the end of the world than the end of capitalism," is also attributed to Fredric Jameson. Adjacent is the notion that the financial crisis was a "crisis of the imagination" as developed by Max Haiven, i.e., the inability to imagine a world outside the global, neoliberal capitalist paradigm; see: Haiven (2014); see also Fisher (2009).

19. Ekbiar and Nardi (2015), 48.

20. The breadth of this work is astonishing and cuts across methodologies. See, among others: Bardzell (2010); Bardzell and Bardzell (2011), (2015); Bødker and Kyng (2018); DiSalvo (2012); Dourish (2017); Dourish and Bell (2011); Feinberg et al. (2019); Forlano (2017); Hansen and Dalsgaard (2015); Irani et al. (2010); Irani and Silberman (2013); Jackson, Gillespie, and Payette (2014); Light (2019); Ligh et al. (2018); Lindtner, Bardzell, and Bardzell (2018); Ratto (2011); Rosner (2018); Rosner, Shorey, Craft, and Remick (2018); Sengers, Boehner, David, and Kaye (2005).

21. Rosner (2018); Sims (2017).

22. Webster (2006).

23. Mitchell (2002); Li (2007); Escobar (1994).

24. Sims (2017).

25. Ferguson (1990); Murphy (2018); Parreñas (2018).

26. Ahmed (2010b).

27. Ibid.

BIBLIOGRAPHY

Abbas, Ackbar. 2000. "Cosmopolitan De-scriptions: Shanghai and Hong Kong." *Public Culture* 12, no. 3: 769–786.

Ahmed, Sara. 2010a. *The Promise of Happiness*. Durham, NC: Duke University Press.

———. 2010b. "Orientations Matter." In *New Materialisms: Ontology, Agency, and Politics*, edited by D. Coole and S. Frost, 234–257. Durham, NC: Duke University Press.

———. 2017. *Living a Feminist Life*. Durham, NC: Duke University Press.

Ames, Morgan. 2016. "Learning Consumption: Media, Literacy, and the Legacy of One Laptop per Child." *The Information Society* 32:2.

———. 2020. *The Charisma Machine: The Life, Death, and Legacy of One Laptop per Child*. Cambridge, MA: MIT Press.

Ames, Morgan, D. Rosner, and I. Erickson. 2015. "Worship, Faith, and Evangelism: Religion as an Ideological Lens for Engineering Worlds." In *Proceedings for CSCW 2015, ACM Conference on Computer-Supported Cooperative Work and Social Computing*, 69–81.

Amrute, Sareeta. 2016. *Encoding Race, Encoding Class: Indian IT Workers in Berlin*. Durham, NC: Duke University Press.

Anagnost, Ann. 1989. "Prosperity and Counter-Prosperity: The Moral Discourse on Wealth in Post-Mao China." In *Marxism and the Chinese Experience: Issues in Contemporary Chinese Socialism*, edited by Arif Dirlik and Maurice Meisner, 210–234. Armonk, NY: M. E. Sharpe.

———. 1997. *National Past-Times: Narrative, Representation, and Power in Modern China*. Durham, NC: Duke University Press.

Anderson, Chris. 2008. "The End of Theory: The Data Deluge Makes the Scientific Method Obsolete." *Wired*. June 23. https://www.wired.com/2008/06/pb-theory/.

———. 2012. *Makers: The New Industrial Revolution*. New York: Crown Business.

Annany, Mike and Kate Crawford. 2016. "Seeing Without Knowing: Limitations of the Transparency Ideal and Its Application to Algorithmic Accountability." *New Media & Society* 20, 3: 973–989.

Aticelli, Tom. 2019. "How Opportunity Zones and Coworking Spaces Joined Forces." *New York Times*, August 20. https://www.nytimes.com/2019/08/20/business/co-working-opportunity-zones.html?auth=login-email&login=email.

Avle, Seyram and Silvia Lindtner. 2016. "Design(ing) Here and There: Tech Entrepreneurs, Global Markets, and Reflexivity in Design Process." *Proceedings of the ACM SIGCHI Conference on Human Factors in Computing Systems, CHI'16*, 2233–2245.

Avle, Seyram, Silvia Lindtner, and Kaiton Williams. 2017. "How Methods Make Designers." In *Proceedings of the 2017 ACM SIGCHI Conference on Human Factors in Computing Systems, CHI'17*, 472–483.

Bach, Jonathan. 2017. "Shenzhen: From Exception to Rule." In *Learning from Shenzhen: China's Post-Mao Experiment from Special Zone to Model City*, edited by Mary Ann O'Donnell, Winnie Wong, and Jonathan Bach, 23–38. Chicago: University of Chicago Press.

Bannon, Liam J., Jeffrey Bardzell, and Suzanne Bødker. 2019. "Reimagining Participatory Design." *Interactions* 26, no. 1: 26–32.

Bannon, Liam J. and Pelle Ehn. 2012. "Design: Design Matters in Participatory Design." In *Routledge International Handbook of Participatory Design*, edited by Jasper Simonsen and Toni Robertson, 37–63. New York: Routledge.

Banzi, Massimo. 2013. "Send in the Clones." Arduino Blog, July 10. https://blog.arduino.cc/2013/07/10/send-in-the-clones/.

Barad, Karen. 2007. *Meeting the Universe Halfway: Quantum Physics and the Entanglement of Matter and Meaning*. Durham, NC: Duke University Press.

Bardzell, Jeffrey. 2018. "Criticality Need Not Be Negative in HCI." *Interaction Culture*. July 30. https://interactionculture.net/2018/07/30/criticality-need-not-be-negative-in-hci/.

Bardzell, Shaowen. 2018. "Utopias of Participation: Feminism, Design, and the Futures." *ACM Transactions on Computer-Human Interaction* 25, no. 1 (February): 1–24.

———. 2010. Feminist HCI: Taking Stock and Outlining an Agenda for Design." In *Proceedings of the ACM SIGCHI Conference on Human Factors in Computing Systems, CHI'10*, 1301–1310.

Bardzell, Shaowen and Jeffrey Bardzell. 2011. "Towards a Feminist HCI Methodology: Social Science, Feminism, and HCI." In *Proceedings of the ACM SIGCHI Conference on Human Factors in Computing Systems, CHI'11*, 675–684.

———. 2015. *Humanistic HCI*. Synthesis Lectures on Human-Centered Informatics. Williston, VT: Morgan and Claypool.

Bardzell, Jeffrey, Shaowen Bardzell, Cindy Lin, Silvia Lindtner, and Autin Toombs. 2017. "HCI's Making Agendas." In *Foundations and Trends® in Human-Computer Interaction* 11, no. 3: 126–200.

Baym, Nancy K. 2015. "Connect with Your Audience! The Relational Labor of Connection." *Communication Review* 18, no. 1: 14–22.

———. 2018. *Playing to the Crowd: Musicians, Audiences, and the Intimate Work of Connection.* New York: New York University Press.

Beck. Eevi E. 2002. "P Is for Political: Participation Is Not Enough." *Scandanavian Journal of Information Systems* 14, no. 1: 77–92.

Beeson, Mark. 2009. "Developmental States in East Asia: A Comparison of the Japanes and Chinese Experiences." *Asian Perspective* 33, no. 2: 5–39.

Bell, Jonathan. 1973. *The Coming of Post-Industrial Society: A Venture in Social Forecasting.* New York: Basic Books.

Benchoff, Brian. 2016. "Arduino vs. Arduino: Arduino Won." *Hackaday*, October 1. https://hackaday.com/2016/10/01/arduino-vs-arduino-arduino-won/.

Benhabib, Seyla. 1992. *Situating the Self: Gender, Community and Postmodernism in Contemporary Ethics*. Cambridge: Polity.

Benjamin, Ruha. 2019. *Race after Technology: Abolitionist Tools for the New Jim Code*. Cambridge, UK: Polity.

Berk, Gerald, Dennis C. Galvan, and Victoria Hattam, eds. 2013. *Political Creativity: Reconfiguring Institutional Order and Change*. Philadelphia: University of Pennsylvania Press.

Berlant, Lauren. 2011. *Cruel Optimism*. Durham, NC: Duke University Press.

Biaglioli, Mario, Peter Jaszi, and Martha Woodmansee, eds. 2011. *Making and Unmaking Intellectual Property: Creative Production in Legal and Cultural Perspective*. Chicago: University of Chicago Press.

Bødker, Susanne. 2006. "When Second Wave HCI Meets Third Way Challenges." In *Proceedings of the 4th Nordic Conference on Human-Computer Interaction: Changing Roles, NorHCI '06*, 1–8.

Bødker, Susanne and Morten Kyng. 2018. "Participatory Design That Matters—Facing the Big Isssues." In *Proceedings of the ACM Transactions on Computer-Human Interaction (TOCHI)*, 25, no. 1, February, article no. 4.

Boellstorff, Tom. 2005. *The Gay Archipelago: Sexuality and Nation in Indonesia*. Princeton, NJ: Princeton University Press.

———. 2008. *Coming of Age in Second Life: An Anthropologist Explores the Virtually Human*. Princeton, NJ: Princeton University Press.

Bolsover, Gillian and Philip Howard. 2018. "Chinese Computational Propaganda: Automation, Algorithms, and the Manipulation of Information about Chinese Politics on Twitter and Weibo." *Information, Communication, and Society* 22, no. 14: 2063–2080.

Boltanski, Luc and Eve Chiapello. 2018. *The New Spirit of Capitalism*, translated by Gregory Elliot. London: Verso.

Booz & Company Inc. (Edward Tse, Kevin Ma, Yu Huang). 2009. "Shan Zhai: A Chinese Phenomenon." Report by Booz & Company Inc.

Bouman, Ole. Public Talk at the *2016 Shenzhen Maker Faire*, Shenzhen, China

boyd, danah, Karen Levy, and Alice Marwick. 2014. "The Networked Nature of Algorithmic Discrimination." *Open Technology Institute | New America | Data & Discrimination* (October): 53–57.

Bray, Francesca. 1997. *Technology and Gender: Fabrics of Power in Late Imperial China*. Cambridge, MA: MIT Press.

Brown, Tim. 2009. *Change by Design: How Design Thinking Transforms Organizations and Insipres Innovation*. New York: Harper Collins.

Brown, Wendy. 2015. *Undoing the Demos: Neoliberalism's Stealth Revolution*. New York: Zone Books.

Buechley, Leah and Benjamin Mako Hill. 2010. "LilyPad in the Wild: How Hardware's Long Tail Is Supporting New Engineering and Design Communities." In *Proceedings of the 8th ACM Conference on Designing Interactive Systems*, 199–207.

Butler, Judith. 1993. *Bodies that Matter: On the Discursive Limits of Sex*. New York: Routledge.

———. 1999. "Gender Is Burning: Questions of Appropriation and Subversion." In *Feminist Film Theory: A Reader*, edited by Sue Thornham, 381–395. New York: New York University Press.

———. 2009. "Performativity, Precarity and Sexual Politics." *AIBR* 4, no. 3: i–xiii.

Calhoun, Craig and Jeffrey N. Wasserstrom. 1999. "Legacies of Radicalism: China's Cultural Revolution and the Democracy Movement of 1989." *Thesis Eleven* 57, no. 1: 33–52.

Callahan, William A. 2004. "Remembering the Future—Utopia, Empire, and Harmony in 21st Century International Theory." *European Journal of International Relations* 10, no. 4 (December): 569–601.

———. 2013. *China Dreams: 20 Visions of the Future*. Oxford: Oxford University Press.

Callon, Michel. 1998. *Laws of the Markets*. Oxford: Blackwell.

Callon, Michel, Yuval Millo, and Fabian Muniesa, eds. 2007. *Market Devices*. Hoboken, NJ: Wiley.

Cartier, Caroyln. 2002. *Globalizing South China*. Malden: Wiley-Blackwell.

———. 2002. "Transnational Urbanism in the Reform-Era Chinese City: Landscapes from Shenzhen." *Urban Studies* 3, no. 9 (August): 1513–1532.

———. 2010. "'Zone Fever,' the Arable Land Debate, and Real Estate Speculation: China's Evolving Land Use Regime and Its Geographical Contradictions." *Journal of Contemporary China* 10, no. 28: 445–469.

Casey, Liam. 2013. "Toronto Teen Wins Thiel Fellowship." *The Star*. Accessed March 2014. https://www.thestar.com/news/gta/2013/05/15/toronto_teen_wins_prestigious_thiel_fellowship.html.

Castells, Manuel. 2009. *The Rise of the Network Society*, 2nd ed. Malden: Wiley-Blackwell.

Chakrabarty, Dipesh. 2007. *Provincializing Europe: Postcolonial Thought and Historical Difference*. Princeton, NJ: Princeton University Press.

Chan, Anita Say. 2014. *Networking Peripheries: Technological Futures and the Myth of Digital Universalism*. Cambridge: MIT Press.

Chen, Nancy N., Constance D. Clark, Suzanne Z. Gottschang, and Lyn Jeffery. 2001. *China Urban: Ethnographies of Contemporary Culture*. Durham, NC: Duke University Press.

Chen, Nancy N. and Lesley A. Sharp. 2014. *Bioinsecurity and Vulnerability*. Santa Fe, NM: School for Advanced Research Press.

Chen, Ting. 2017. *A State Beyond the State: Shenzhen and the Transformation of Urban China*. Rotterdam: nai010 publishers.

Chen, Xiangming and Taylor Lynch Ogan. 2017. "China's Emerging Silicon Valley: How and Why Has Shenzhen Become a Global Innovation Centre." *European Financial Review* (December–January): 55–62.

Christenson, Clayton M. 1997. *The Innovator's Dilemma: The Revolutionary Book That Will Change the Way You Do Business*. Boston: Harvard Business School Press.

Chu, Nellie. 2016. "The Emergence of 'Craft' and Migrant Entrepreneurship Along the Global Commodity Chains for Fast Fashion in Southern China." *Journal of Modern Craft* 9, no. 2: 193–213.

Chumley, Lily. 2016. *Creativity Class: Art School and Culture Work in Postsocialist China*. Princeton, NJ: Princeton University Press.

Chun, Wendy Hui Kyong. 2006. *Control and Freedom: Power and Paranoia in the Age of Fiber Optics*. Cambridge, MA: MIT Press.

———. 2016. *Updating to Remain the Same: Habitual New Media*. Cambridge, MA: MIT Press.

Clay, Kelly. 2013. "3D Printing Company MakerBot Acquired in $604 Million Deal." *Forbes*, June 19, https://www.forbes.com/sites/kellyclay/2013/06/19/3d-printing-company -makerbot-acquired-in-604-million-deal/#4f8442cc1ef8.

Cleaver, Harry. 1992. "The Inversion of Class Perspective in Marxian Theory: From Valorization to Self-Valorization." In *Open Marxism*, vol. 2, edited by Werner Bonefeld, Richard Gunn, and Kosmos Psychopedis, 106–144. London: Pluto.

Cockburn, Cynthia. 1991. *In the Way of Women: Men's Resistance to Sex Equality in Organizations*. Ithaca, NY: ILR Press.

Cockayne, Daniel G. 2015. "Entrepreneurial Affecet: Attachment to Work Practice in San Francisco's Digital Media Sector." *Environment and Planning D: Society and Space* 34(3): 456–473.

Cohn, Marisa. 2016. "Convivial Decay: Entangled Litetimes in a Geriatric Infrastructure." In *Proceedings of the 19th ACM Conference on Computer Supported Cooperative Work and Social Computing (CSCW'16)*, San Francisco, CA, 1511–1523.

Cole, Alyson and Victoria Hattam, eds. 2017. "Precarious Work." *Women Studies Quarterly* 45, nos. 3, 4.

Coleman, E. Gabriella. 2013. *Coding Freedom: The Ethics and Aesthetics of Hacking*. Princeton, NJ: Princeton University Press.

Collier, Stephen. 2011. *Post-Soviet Social: Neoliberalism, Social Modernity, Biopolitics*. Princeton, NJ: Princeton University Press.

Comaroff, Jean. 1985. *Body of Power, Spirit of Resistance: The Culture and History of a South African People*. Chicago: University of Chicago Press.

Constanza-Chock, Sasha. 2020. *Design Justice: Community-Led Practices to Build the World We Need*. Cambridge, MA: MIT Press.

Cooper, Melinda and Catherine Waldby. 2014. *Clinical Labor: Tissue Donors and Research Subjects in the Global Bioeconomy*. Durham, NC: Duke University Press.

Cormier, Brendan. 2015. "Researching Unidentified Acts of Design." *V&A Blog*, October 14. https://www.vam.ac.uk/blog/international-initiatives/unidentified-acts-of-design-at-the -uabb.

Cowen, Deborah. 2014. *The Deadly Life of Logistics: Mapping Violence in Global Trade*. Minneapolis: University of Minnesota Press.

Croll, Elizabeth. 1995. *Changing Identities of Chinese Women: Rhetoric, Experience and Self-Perception in 20th Century China*. Hong Kong: Hong Kong University Press.

Cross, Jamie. 2014. *Dream Zones: Anticipating Capitalism and Development in India*. London: Pluto Press.

Cull, Nikolas. 2012. "The Legacy of the Shanghai Expo and Chinese Public Diplomacy." *Place Branding and Public Diplomacy* 8, no. 2: 99–101.

Curtis, Michael and Kevin Sanson, eds. 2016. *Precarious Creativity: Global Media, Local Labor*. Oakland: University of California Press.

Davis, Gerald F. 2016a. *The Vanishing American Corporation: Navigating the Hazards of a New Economy*. Oakland, CA: Berret-Koehler.

———. 2016b. "Can an Economy Survive Without Corporations? Technology and Robust Organizational Alternatives." *Academy of Management Perspectives* 30, no. 2: 129–140.

De Bary, Theodore and Richard Lufrano. 2000. *Sources of Chinese Tradition, Vol. II: From 1600 Through the Twentieth Century*. New York: Columbia University Press.

Deuze Mark. 2007. *Media Work*. Cambridge: Polity.

Devendorf, Laura and Kimiko Ryokai. 2014. "Being the Machine: Exploring New Modes of Making." In *Proceedings of the Design Interactive Systems Conference*, 33–36.

Dirlik, Arif. 1989. "Revolutionary Hegemony and the Language of Revolution: Chinese Socialism Between Present and Fuutre." In *Marxism and the Chinese Experience: Issues in Contemporary Chinese Socialism*, edited by Arif Dirlik and Maurice J. Meisner. Armonk, NY: M. E. Sharpe, 27–39.

———. 1997. *The Postcolonial Aura: Third World Criticism in the Age of Global Capitalism*. Boulder, CO: Westview Press.

———. 2012a. "The Idea of a 'Chinese Model': A Critical Discussion." *China Information* 26, no. 3: 277–302.

———. 2012b. *Culture and History in Postrevolutionary China: The Perspective of Global Modernity*. Hong Kong: Chinese University Press.

Dirlik, Arif and Maurice J. Meisner. 1989. *Marxism and the Chinese Experience: Issues in Contemporary Chinese Socialism*. Armonk, NY: M. E. Sharpe.

DiSalvo, Carl. 2012. *Adversarial Design*. Cambridge, MA: MIT Press.

Dong, Yige. 2019. *From Mill Town to iPhone City: Gender, Labor, and the Politics of Care in an Industrializing China (1949–2017)*. PhD Diss., Johns Hopkins University, Baltimore, MD.

Dougherty, Dale. 2015. "The Future of Work: Join the Maker Movement." *Pacific Standard*, November 5.

Douglas, Thomas. 2002. *Hacker Culture*. Minneapolis: University of Minnesota Press.

Dourish, Paul. 2007. "Seeing Like an Interface." In *Proceedings of the Australian Computer-Human Interaction Conference OzCHI*, 1–8. Adelaide, Australia.

———. 2017. *The Stuff of Bits: An Essay on the Materialities of Information*. Cambridge, MA: MIT Press.

Dourish, Paul and Genevieve Bell. 2011. *Divining a Digital Future: Mess and Mythology in Ubiquitous Computing*. Cambridge, MA: MIT Press.

———. 2018. "'Resistance Is Futile': Reading Science Fiction Alongside Ubiquitous Computing." *Personal and Ubiquitous Computing* 18, no. 4 (April): 769–778.

Dourish, Paul and Scott Mainwaring. 2012. "Ubicomp's Colonial Impulse." In *Proceedings of the 2012 ACM Conference on Ubiquitous Computing*, 133–142.

Duffy, Brooke Erin. 2017. *(Not) Getting Paid to Do What You Love: Gender, Social Media, and Aspirational Work*. New Haven, CT: Yale University Press.

Dunbar-Hester, Christina. 2014. *Low Power to the People: Pirates, Protest, and Politics in FM Radio Activism*. Cambridge, MA: MIT Press.

Dunbar-Hester, Christina. 2020. *Hacking Diversity: The Politics of Inclusion in Open Technology Cultures*. Princeton, NJ: Princeton University Press.

Dyer-Witheford, Nick. 1999. *Cyber-Marx: Cycles and Circuits of Struggle in High Technology Capitalism*. Urbana: University of Illinois Press.

Easterling, Keller. 2014. *Extrastatecraft: The Power of Infrastructure Space*. Brooklyn, NY: Verso.

Edwards, Paul. 1996. *The Closed World: Computers and the Politics of Discourse in Cold War America*. Cambridge, MA: MIT Press.

Eglash, Ron. 2002. "Race, Sex, and Nerds: From Black Geeks to Asian American Hipsters." *Social Text* 72, vol. 20, no. 2: 49–64.

———. 2016. "Of Marx and Makers: An Historical Perspective on Generative Justice." *Teknokultura* 13, no. 1: 245–269.

Ehn, Pelle, Elizabet M. Nilsson, and Richard Topgaard, eds. 2014. *Making Futures: Marginal Notes on Innovation, Design, and Democracy*. Cambridge, MA: MIT Press.

Ekbiar, Hamid and Bonnie Nardi. 2015. "The Political Economy of Computing: The Elephant in the HCI Room." *Interactions Magazine*, 22.6, November + December, 46.

Elyachar, Julia. 2005. *Markets of Dispossession: NGOS, Economic Development, and the State in Cairo*. Durham, NC: Duke University Press.

Ensmenger, Nathan L. 2003. "Letting the 'Computer Boys' Take Over: Technology and the Politics of Organizational Transformation." *International Review of Social History* 48, no. S11 (December): 153–180.

———. 2010. *The Computer Boys Take Over: Computers, Programmers, and the Politics of Technical Expertise*. Cambridge, MA: MIT Press.

Escobar, Arturo. 1994. *Encountering Development: The Making and Unmaking of the Third World*. Princeton, NJ: Princeton University Press.

———. 2017. *Designs for the Pluriverse: Radical Interdependence, Autonomy, and the Making of Worlds*. Durham, NC: Duke University Press.

Etherington, Darrell. 2013. "Stratays Acquiring MakerBot in $403M Deal, Combined Company Will Likely Domiante 3D Printing Industry." *Techcrunch*, June 19, https://techcrunch.com/2013/06/19/stratasys-acquiring-makerbot-combined-company-will-likely-dominate-3d-printing-industry/.

Eubanks, Virginia. 2018. *Automating Inequality: How High-Tech Tools Profile, Police, and Punish the Poor*. New York: St. Martin's Press.

Faubion, James D. and George E. Marcus. 2009. *Fieldwork Is Not What It Used to Be: Learning Anthropology's Method in a Time of Transition*. Ithaca, NY: Cornell University Press.

Feher, Michel. 2009. "Self-Appreciation; or, The Aspirations of Human Capital." *Public Culture* 21:1.

———. 2018. *Rated Agency: Investee Politics in a Speculative Age*. New York: Zone Books.

Feinberg, Melanie, Sarah Fox, Jean Hardy, Stephanie Steinhardt, and Palashi Vaghela. 2019. "At the Intersection of Culture and Method: Designing Feminist Action." In *Companion Publication of the 2019 on Designing Interactive Systems (DIS'19) Conference*, 365–368.

Ferguson, James. 1994. *The Anti-Politics Machine: "Development," Depoliticization, and Bureaucratic Power in Lesotho*. Minneapolis, MN: University of Minnesota Press.

———. 2006. *Global Shadows: Africa in the Neoliberal World Order*. Durham, NC: Duke University Press.

Fewsmith, Joseph. 2008. *China Since Tiananmen*. Cambridge: Cambridge University Press.

Fine, Gary Alan. 1979. "Small Groups and Culture Creation: The Idioculture of Little League Baseball Teams." *American Sociological Review* 44: 733–745.

Fisher, Anna Watkins. 2020. *The Play in the System: The Art of Parasitical Resistance*. Durham, NC: Duke University Press.

Fisher, Mark. 2009. *Capitalist Realism: Is There No Alternative? Hants, UK:* Zero Books, John Hunt Publishing.

Flew, Terry and Stuart Cunningham. 2010. "Creative Industries after the First Decade of Debate." *Information Society* 26, no. 2: 113–123.

Florence, Eric. 2017. "How to Be a Shenzhener: Representations of Migrant Labor in Shenzhen's Second Decade." In *Learning from Shenzhen: China's Post-Mao Experiment from Special Zone to Model City*, edited by Mary Ann O'Donnell, Winnie Wong, and Jonathan Bach, 86–103. Chicago: University of Chicago Press.

Florida, Richard. 2002. *The Rise of the Creative Class: And How It's Transforming Work, Leisure, Community and Everyday Life*. New York: Basic Books.

Forlano, Laura. 2017. "Posthumanism and Design." *Sheji—the Journal of Design, Economics, and Innovation* 3, no. 1 (Spring): 16–29.

Forlano, Laura and Megan Halpern. 2015. "Reimagining Work: Entanglements and Frictions around Future of Work Narratives." *Fibreculture (26), Special Issue on "Entanglements: Activism and Technology* Issn: 1149–1443, 32–59.

Foucault, Michel. 1980. *Discipline and Punish: The Birth of the Prison*. New York: Pantheon.

———. 1994. *The Order of Things: An Archeology of the Human Sciences (1966)*. New York: Vintage.

———. 2008. *The Birth of Biopolitics: Lectures at the Collége de France, 1978–1979*. Translated by Graham Burchell. New York: Palgrave Macmillan.

Fox, Sarah, Rachel Rose Ulgado, and Daniela Rosner. 2015. "Hacking Culture, Not Devices: Access and Recognition in Feminist Hackerspaces." In *Proceedings of the 18th ACM Conference on Computer Supported Cooperative Work & Social Computing*, 56–68. ACM, Vancouver, BC.

Freeman, Guo, Jeffrey Bardzell, and Shaowen Bardzell. 2017. "Aspirational Design and Messy Democracy: Partisanship, Policy, and Hope in an Asian City." In *Proceedings of the ACM Conference on Computer Supported Cooperative Work & Social Computing*. ACM: New York.

Fuchs, Christian. 2014. *Digital Labour and Karl Marx*. New York: Routledge.

Gallagher, Mary E. 2005. *Contagious Capitalism: Globalization and the Politics of Labor in China*. Princeton, NJ: Princeton University Press.

Geier, Ben. 2016. "Even the IMF Now Admits Neoliberalism Has Failed." *Fortune*, June 3. https://fortune.com/2016/06/03/imf-neoliberalism-failing/.

Geron, Tomio. 2002. "Building on Maker Movement, Hardware Startups Pitch at HAXLR8R Demo Day." *Forbes*, June 18. https://www.forbes.com/sites/tomiogeron/2012/06/18/building-on-maker-movement-hardware-startups-pitch-at-haxlr8r-demo-day/#3535274b5513.

Gibson-Graham, J. K. 2006. *A Postcapitalist Politics*. Minneapolis: University of Minnesota Press.

Gillespie, Tarleton. 2009. "Characterizing Copyright in the Classroom: The Cultural Work of Antipiracy Campaigns." *Communication, Culture, & Critique* 2, no. 3 (September): 274–318.

———. 2014. "The Relevance of Algorithms." In *Media Technologies: Essays on Communication, Materiality, and Society*, edited by Tarleton Gillespie, Pablo J. Boczkowski, and Kirsten A. Foot, 167–193. Cambridge, MA: MIT Press.

Gray, Mary L. and Siddharth Suri. 2019. *Ghost Work: How to Stop Silicon Valley from Building a New Global Underclass. New York:* Eamon Dolan/Houghton Mifflin Harcourt.

Greenhalgh, Susan. 2010. *Cultivating Global Citizens: Population in the Rise of China*. Cambridge, MA: Harvard University Press.

Greenhalgh, Susan and Edwin A. Winckler. 2005. *Governing China's Population: From Leninist to Neoliberal Biopolitics*. Stanford, CA: Stanford University Press.

Greenspan, Anna. 2014. *Shanghai Future: Modernity Remade*. New York: Oxford University Press.

Gregg, Melissa. 2009. "Learning to (Love) Labour: Production Cultures and the Affective Turn." *Communication and Critical/Cultural Studies* 6, no. 2: 209–214.

———. 2011. *Work's Intimacy*. New York: Polity.

Gregg, Melissa and Gregory J. Seigworth. 2010. *The Affect Theory Reader*. Durham, NC: Duke University Press.

Haiven, Max. 2014. *Crises of Imagination, Crises of Power: Capitalism, Culture and Resistance in a Post-Crash World*. London: Zed Books.

Hall, Gary. 2016. *The Uberfication of the University*. Minneapolis: University of Minnesota Press.

Hansen, Lone Koefoed and Peter Dalsgaard. 2015. "Note to Self: Stop Calling Interfaces 'Natural.'" In *Proceedings of the Fifth Decennial Aarhus Conference on Critical Alternatives*, August, 65–68.

Hanser, Amy. 2008. *Service Encounters: Class, Gender, and the Market for Social Distinction in Urban China*. Stanford, CA: Stanford University Press.

Haraway, Donna. 1991. *Simians, Cyborgs, and Women: The Reinvention of Nature*. New York: Routledge.

"Hardware Hackers Aren't Waiting for Self-Driving Cars to Hit the Road." 2017. YouTube video, 4:28. Posted by "Make," February 9. https://www.youtube.com/watch?v=xjT-UykifDc.

Harney, Stefano and Fred Moten. 2013. *The Undercommons: Fugitive Planning & Black Study*. Brooklyn, NY: Autonomedia.

Harold, David Kurt and Peter Marolt, eds. 2011. *Online Society in China: Creating, Celebrating, and Instrumentalizing the Online Carnival*. London: Routledge.

Harvey, David. 1990. *The Condition of Postmodernity: An Inquiry into the Origins of Cultural Change*. Malden, MA: Blackwell.

———. 2000. *Spaces of Hope*. Berkeley: University of California Press, 2000.

———. 2007a. *A Brief History of Neoliberalism*. Oxford: Oxford University Press.

———. 2007b. *Limits to Capital*. London: Verso.

———. 2008. "The Right to the City." *New Left Review* 53 (September–October): 23–40.

Hayden, Cori. 2010. "The Proper Copy: The Insides and Outsides of Domains Made Public." *Journal of Cultural Economy* 3, no. 1 (March): 85–102.

Hayden, Craig. 2011. *The Rhetoric of Soft Power: Public Diplomacy in Global Contexts*. Lanham, MD: Lexington Books.

Hecht, Gabrielle. 1998. *The Radiance of France: Nuclear Power and National Identity after World War II*. Cambridge, MA: MIT Press.

———. 2002. "Rupture Talk in the Nuclear Age: Conjugating Colonial Power in Africa." *Social Studies of Science* 32, nos. 5–6.

———. 2012. *Being Nuclear: Africans and the Global Uranium Trade*. Cambridge, MA: MIT Press.

———. 2018. "Interscalar Vehicles for an African Anthropocene: On Waste, Temporality, and Violence." *Cultural Anthropology* 33, no. 1: 109–141.

Heilmann, Sebastian. 2008. "From Local Experiments to National Policy: The Origins of China's Distinctive Policy Process." *China Journal* 59: 1–30.

———.. 2016. "Leninism Upgraded: Xi Jinping's Authoritarian Innovations." *China Economic Quarterly* 20, no. 4: 15–22.

Heilmann, Sebastian and Elizabeth J. Perry, eds. 2011. *Mao's Invisible Hand: The Political Foundations of Adaptive Governance in China*. Cambridge, MA: Harvard University Press.

Hershatter, Gail. 2007. *Women in China's Long Twentieth Century*. Berkeley: University of California Press.

Hertz, Garnet. 2014. "Critical Making." Concept Lab. http://www.conceptlab.com/criticalmaking/.

Hicks, Marie. 2018. *Programmed Inequality: How Britain Discarded Women Technologists and Lost Its Edge in Computing*. Cambridge, MA: MIT Press.

Hird, Derek. 2009. "White-Collar Men and Masculinities in Contemporary China." PhD Diss., University of Westminster School of Social Sciences, Humanities and Languages.

Hird, Derek and Song, Geng. 2019. *The Cosmopolitan Dream: Transnational Chinese Masculinities in a Global Age*. Hong Kong: Hong Kong University Press.

Hirosue, Sachiko, Denisa Kera, and Hermes Huang. 2015. "Promises and Perils of Open Source Technologies for Development: Can the 'Subaltern' Research and Innovate?" In *Technologies for Development*, edited by S. Hostettler, E. Hazboun, and J. C. Bolay, 73–80. Berlin: Springer.

Ho, Josephine. 2010. "ShanZhai: Economic/Cultural Production Through the Cracks of Globalization." Plenary paper presented at ACS Crossroads: 2010 Cultural Studies Conference, Hong Kong, June. http://sex.ncu.edu.tw/members/Ho/20100617%20Crossroads%20Plenary%20 Speech.pdf.

Hochschild, Arlie Russell. 1983. *The Managed Heart: Commercialization of Human Feeling*. Berkeley: University of California Press.

———. 2018. "The Concept Creep of 'Emotional Labor.'" In *An Interview with Julie Beck: The Atlantic*, November 26. https://www.theatlantic.com/family/archive/2018/11/arlie-hochschild -housework-isnt-emotional-labor/576637/.

Hoffman, Lisa M. 2010. *Patriotic Professionalism in Urban China: Fostering Talent*. Philadelphia: Temple University Press.

hooks, bell. 1990. *Yearning: Race, Gender, and Cultural Politics*. Boston: South End Press.

Howard, Philip. 2015. *Pax Technica: How the Internet of Things May Set Us Free or Lock Us Up*. New Haven, CT: Yale University Press.

Huang, Andrew "bunnie." 2013a. "The $12 'Gongkai' Phone." *Bunnie Studios*. https://www .bunniestudios.com/blog/?page_id=3107.

———. 2013b. "Why the Best Days of Open Hardware Are Yet to Come." Public Talk. *Generator Conference*, Shenzhen, Guangdong, China, January 26.

———. 2013c. *Hacking the Xbox: An Introduction to Reverse Engineering*. San Francisco, CA: No Starch Press.

———. 2014. "From Gongkai to Open Source." *Bunnie Studios*. https://www.bunniestudios.com /blog/?p=4297.

Huang, Weiwen. 2017. "The Tripartate Origins of Shenzhen: Beijing, Hong Kong, and Bao'an." In *Learning from Shenzhen: China's Post-Mao Experiment from Special Zone to Model City*, edited by Mary Ann O'Donnell, Winnie Wong, and Jonathan Bach, 65–85. Chicago: University of Chicago Press.

Huang, Yu and Naubahar Sharif. 2017. "From 'Labor Dividend' to 'Robot Dividend': Technological Change and Workers' Power in South China." *Agrarian South: Journal of Political Economy* 6, no. 1: 53–78.

Hubbert, Jennifer. 2017. "Back to the Future: The Politics of Culture at the Shanghai Expo." *International Journal of Cultural Studies* 20, no. 1: 48–64.

Hughes, Bob. 2005. "From Useful Idiocy to Activism: A Marxist Interpretation of Computer Development." AARHUS'05, ACM, Aarhus, Denmark.

Hui, Wang. 2006. *China's New Order: Soceity, Economics, and Politics in Transition*. Edited and translated by Theodore Huters. Cambridge, MA: Harvard University Press.

———. 2010. The Economy of Rising China and Its Contradictions." *Reading the China Dream*. https://www.readingthechinadream.com/wang-hui-the-economy-of-rising-china.html; based on: 汪晖, "中国崛起的经验及其面临的挑战," 文化纵横 (*Beijing Cultural Review*) 2: 24–35.

———. 2011. *The End of the Revolution: China and the Limits of Modernity*. London: Verso.

Huppatz, D. J. 2015. "Revisiting Herbert Simon's 'Science of Design.'" *Design Issues* 31, no. 2 (Spring): 29–40.

IDEO Patterns. 2009–2011. *Shanzhai*. Issue 27. https://www.ideo.com/post/ideo-patterns.

Irani, Lilly. 2013. "The Cultural Work of Microwork." *New Media & Society* 17, no. 5: 720–739.

———. 2015. "Difference and Dependence among Ddigital Workers: The Case of Amazon Mechanical Turk." *South Atlantic Quarterly* 114, no. 4: 225–234.

———. 2018. "'Design Thinking': Defending Silicon Valley at the Apex of Global Labor Heirarchies." *Catalyst* 4, no. 1: 1–19.

Irani, Lilly. 2019. *Chasing Innovation: Making Entrepreneurial Citizens in Modern India.* Princeton, NJ: Princeton University Press.

Irani, Lilly and Kavita Philip. 2018. "Negotiating Engines of Difference." *Catalyst: Feminism, Theory, Technoscience* 4, no. 2: 1–11.

Irani, Lilly and Michael "Six" Silberman. 2013. "Turkopticon: Interrupting Worker Invisibility in Amazon Mechanical Turk." In *Proceedings of the SIGCHI Conference on Human Factors in Computing Systems* (CHI '13), 611–620.

Irani, Lilly, Janet Vertesi, Paul Dourish, Kavita Philip, and Rebecca Grinter. 2010. "Postcolonial Computing: A Lens on Design and Development." In *Proceedings of the SIGCHI Conference on Human Factors in Computing Systems (CHI 2010)*, 1311–1320.

IRS. "Opportunity Zones Frequently Asked Questions." https://www.irs.gov/newsroom/opportunity-zones-frequently-asked-questions#qof.

Iskander, Natasha. 2018. Design Thinking Is Fundamentally Conservative and Preserves the Status Quo." *Harvard Business Review*, September. https://hbr.org/2018/09/design-thinking-is-fundamentally-conservative-and-preserves-the-status-quo.

Ito, Joi. 2014a. "Shenzhen Trip Report—Visiting the World's Manufacturing Ecosystem." Joi Ito Weblog, September 1, https://joi.ito.com/weblog/2014/09/01/shenzhen-trip-r.html.

———. 2014b. "Want to Innovate? Become a 'Now-ist.'" TED2014, March. Accessed April 2014, https://www.ted.com/talks/joi_ito_want_to_innovate_become_a_now_ist?language=en.

———. 2019. "Supposedly 'Fair' Algorithms Can Perpetuate Discrimination." *Wired* magazine, May, https://www.wired.com/story/ideas-joi-ito-insurance-algorithms/.

Ito, Mizuko. 2009. *Engineering Play: A Cultural History of Children's Software.* Cambridge, MA: MIT Press.

Jackson, Steven J. 2014. "Rethinking Repair." In *Media Technologies: Essays on Communication, Materiality and Society.* Cambridge, MA: MIT Press.

Jackson, Steven J., Tarleton Gillespie, and Sandra Payette. 2014. "The Policy Knot: Reintegrating Policy, Practice and Design in CSCW Studies of Social Computing." In *Proceedings of the 17th ACM Conference on Computer Supported Cooperative Work and Social Computing (CSCW'14)*, 588–602.

Jeffery, Lyn. 2014. "Mining Innovation from an Unexpected Source: Lessons from the Shanzhai." Lecture at the Shenzhen Maker Faire.

Jen, Natasha. 2018. "Design Thinking Is B.S." *Fast Company*, September. https://www.fastcompany.com/90166804/design-thinking-is-b-s.

Johnson, Chalmers A. 1982. *MITI and the Japanese Miracle: The Growth of Industrial Policy, 1925–1975.* Stanford, CA: Stanford University Press.

Jones, Andrew F. 2011. *Developmental Fairy Tales: Evolutionary Thinking and Modern Chinese Culture.* Cambridge, MA: Harvard University Press.

Joseph, Ralina L. 2018. *Postracial Resistance: Black Women, Media, and the Uses of Strategic Ambiguity.* New York: NYU Press.

Justice, Lorraine. 2012. *China's Design Revolution.* Cambridge, MA: MIT Press.

Kanstrup, Anne Marie. 2003. "D Is for Democracy: On Political Ideals in Participatory Design." *Scandinavian Journal of Information Systems* 15, no. 1: 81–85.

Kaziunas, Elizabeth, Mark Ackerman, Silvia Lindtner, and Joyce Lee. 2017. "Caring through Data: Attending to the Social and Emotional Experiences in Health Datafication." In *Proceedings of the 2017 ACM Conference on Computer Supported Cooperative Work and Social Computing, CSCW 2017*, 2260–2272.

Kaziunas, Elizabeth, Silvia Lindtner, Mark Ackerman, and Joyce Lee. 2017. "Lived Data: Tinkering with Bodies, Code, and Care Work." *Journal of Human-Computer Interaction* 33, no. 1: 49–92.

Keane, Michael. 2011. *China's New Creative Clusters: Governance, Human Capital, and Investment.* London: Routledge.

———. 2013. *Creative Industries in China: Art, Design, and Media.* Cambridge: Polity.

———, ed. 2016. *Handbook of Cultural and Creative Industries in China.* Cheltenham: Edward Elgar.

Kelty, Christopher. 2008. *Two Bits: The Cultural Significance of Free Software.* Durham, NC: Duke University Press.

Kensing, Finn and Jeanette Blomberg. 1998. "Participatory Design: Issues and Concerns." *Journal of Computer Supported Cooperative Work* 7 nos. 3–4: 167–185.

Kipnis, Andrew, ed. 2012. *Chinese Modernity and the Individual Psyche.* New York: Palgrave Macmillan.

Kleif, Tine and Wendy Faulkner. 2003. "'I'm No Athlete [but] I Can Make This Thing Dance!'— Men's Pleasures in Technology." *Science, Technology, & Human Values* 28, no. 2: 296–325.

Knorr Cetina, Karin. 1999. *Epistemic Cultures: How the Sciences Make Knowledge.* Cambridge, MA: Harvard University Press.

Kokas, Aynne. 2017. *Hollywood Made in China.* Oakland: University of California Press.

Kolko, Jon. 2018. "The Divisiveness of Design Thinking." *Interactions Magazine*, 25.3, May–June.

Kreiss, Daniel. 2016. *Prototype Politics: Technology-Intensive Campaigning and the Data of Democracy.* Oxford: Oxford University Press.

Kreiss, Daniel, Megan Finn, and Fred Turner. 2011. "The Limits of Peer Production: Some Reminders from Max Weber for the Network Society." *New Media & Society* 13, no. 2: 243–259.

Kroeber, Arthur R. 2017. *China's Economy: What Everyone Needs to Know*°. Oxford: Oxford University Press.

Kücklich, Julian. 2005. "Precarious Playbour: Modders and the Digital Games Industry." *Fibreculture Journal*, Issue 5: Precarious Labor.

Kuruvilla, Sarosh, Ching Kwan Lee, and Mary E. Gallagher. 2011. *From Iron Rice Bowl to Informalization: Markets, Workers, and the State in Changing China.* Ithaca, NY: ILR Press.

Kuznetsov, Stacey and Eric Paulos. 2010. Rise of the Expert Amateur. In *Proceedings of the 6th Nordic Conference on Human-Computer Interaction Extending Boundaries—NordiCHI '10.* New York: ACM Press, 295.

Kuznetsov, Stacey, Nathan Wilson, Scott Hudson, Carrie Doonan, Swarna Mohan, and Eric Paulos. 2015. "DIYBio Things: Open Source Biology Tools as Platforms for Hybrid Knowledge Production and Scientific Participation." In *Proceedings of the ACM SIGCHI Conference on Human Factors in Computing Systems*, 4065–4068.

Kyng, Martin. 2010. "Bridging the Gap Between Politics and Techniques: On the Next Practices of Participatory Design." *Scandinavian Journal of Information Systems* 22, no. 1: 49–68.

La Repubblica. 2015. "Massimo Banzi: 'Il perché della guerra per Arduino.'" February 11. http://playground.blogautore.repubblica.it/2015/02/11/la-guerra-per-arduino-la-perla-hi-tech-italiana-nel-caos/?refresh_ce.

Latimer, Joanna. 2013. "Being Alongside: Rethinking Relations amongst Different Kinds." *Theory, Culture & Society* 30, no. 7–8 (December): 77–104.

Latour, Bruno. 2004. "Why Has Critique Run Out of Steam? From Matters of Fact to Matters of Concern." *Critical Inquiry* 30, no. 2 (Winter): 225–248.

———. 2005. *Reassembling the Social: An Introduction to Actor-Network-Theory.* Oxford: Oxford University Press.

Latour, Bruno and Steve Woolgar. 1986. *Laboratory Life: The Construction of Scientific Facts.* Princeton, NJ: Princeton University Press.

Lazzarato, Maurizio. 1996. "Immaterial Labor." In *Radical Thought in Italy,* edited by Paolo Virno and Michael Hardt, 133–146. Minneapolis: University of Minnesota Press.

Lee, Ching Kwan. 1998. *Gender and the South China Miracle: The Worlds of Factory Women*. Berkeley: University of California Press.

———. 2007. *Against the Law: Labor Protest in China's Rustbelt and Sunbelt*. Berkeley: University of California Press.

Lee, Leo Ou-fan. 1999. *Shanghai Modern: The Flowering of a New Urban Culture in Shanghai, 1930–1945*. Cambridge, MA: Harvard University Press.

Lessig, Lawrence. 1999. *Code: And Other Laws of Cyberspace*. New York: Basic Books.

———. 2005. *Free Culture: The Nature and Future of Creativity*. London: Penguin Books.

———. 2008. *Remix: Making Art and Commerce Thrive in the Hybrid Economy*. New York: Penguin.

Levien, Michael. 2018. *Dispossession Without Development: Land Grabs in Neoliberal India*. New York: Oxford University Press.

Levitas, Ruth. 2013. *Utopia as Method: The Imaginary Reconstitution of Society*. New York: Palgrave Macmillan.

Levy, Steven. 2010. *Hackers: Heroes of the Computer Revolution*. Newton: O'Reilly Media.

Li, Tanya. 2007. *The Will to Improve: Governmentality, Development, and the Practice of Politics*. Durham, NC: Duke University Press.

Liang, Lawrence. 2014. "Beyond Representation: The Figure of the Pirate." In *Postcolonial Piracy: Media Distribution and Cultural Production in the Global South*, edited by Lars Eckstein and Anja Schwarz, 49–77. London: Bloomsbury.

Light, Ann. 2019. "Design and Social Innovation at the Margins: Finding and Making Cultures of Plurality." *Journal of Design and Culture* 11, no. 1: 13–35.

Light, Ann, Chris Frauenberger, Jennifer Preece, Paul Strohmeier, and Maria Angela Ferrario. 2018. "Taking Action in a Changing World." *Interactions Magazine* 25, no. 1 (January + February): 34–45.

Lin, Cindy Kaiying, Silvia Lindtner, and Stefanie Wuschitz. 2019. "Hacking Difference in Indonesia: The Ambivalences of Designing for Alternative Futures." In *Proceedings of ACM Conference on Designing Interactive Systems, DIS 2019*, 1571–1582.

Lin, Yuwei. 2005. "Gender Dimensions of FLOSS Development." *Mute* 2, no. 1: 38–42.

Lindtner, Silvia. 2012. *Cultivating Creative China: The Making and Remaking of Cities, Citizens, Work, and Innovation*. UMI Dissertation Publishing, Proquest.

———. 2015. "Hacking with Chinese Characteristics: The Promises of the Maker Movement Against China's Manufacturing Culture." *Science, Technology, & Human Values* 40, no. 5: 854–879.

———. 2017. "Laboratory of the Precarious: The Seductive Draw of Entrepreneurial Living." *Women's Studies Quarterly* 45, nos. 3 and 4: 287–305.

Lindtner, Silvia, Shaowen Bardzell, and Jeffrey Bardzell. 2016. "Reconstituting the Utopian Vision of Making: HCI After Technosolutionism." In *Proceedings of the ACM SIGCHI Conference on Human Factors in Computing Systems*, 1390–1402.

———. 2018. "Design and Intervention in the Age of 'No Alternative.'" In *Proceedings of the ACM Conference on Human-Computer Interaction—CSCW* 2, no. 109 (November): 1–21.

Lindtner, Silvia, Anna Greenspan, and David Li. 2015. "Designed in Shenzhen: Shanzhai Manufacturing and Maker Entrepreneurs." In *Proceedings of the Fifth Decennial Aarhus Conference in Critical Alternatives*, 85–96. Aarhus: Aarhus University Press.

Lindtner, Silvia and Cindy Lin Kaiying. 2017. "Making and Its Promises." *Journal of CoDesign* 2: 70–82.

Lindtner, Silvia and Marcella Szablewicz. 2011. "China's Many Internets: Participation and Digital Game Play Across a Changing Technology Landscape." In *Online Society in China: Creating, Celebrating, and Instrumentalizing the Online Carnival*, edited by David Kurt Harold and Peter Marolt, Routledge, 89–105. London: Routledge.

Ling, L.H.M. 1999. "Sex Machine: Global Hypermasculinity and Images of Asian Women in Modernity." In *Positions: East Asia Cultures Critique* 7: 277–306.

Litzinger, Ralph A. 1998. "Memory Work: Reconstituting the Ethnic in Post-Mao China." *Cultural Anthropology* 13, no. 2 (May): 224–255.

———. 2016. "Regimes of Exclusion and Inclusion: Migrant Labor, Education, and Contested Futurities." In *Ghost Protocol: Development and Displacement in Global China*, edited by Carlos Rojas and Ralph A. Litzinger, 191–204. Durham, NC: Duke University Press.

Lorde, Audre. 2007. "The Master's Tools Will Never Dismantle the Master's House," *Sister Outsider: Essays and Speeches*. Berkeley, CA: Crossing Press.

Lorey, Isabell. 2010. "Becoming Common: Precarization as Political Constituting." E-flux, Journal #17—June.

———. 2015. *State of Insecurity: Government of the Precarious*. New York: Verso.

Lu, Xiaobo and Elizabeth J. Perry. 2015. *Danwei: The Changing Chinese Workplace in Comparative and Historical Perspective*. London: Routledge.

MacFarquhar, Roderick and Michael Schoenhals. 2006. *Mao's Last Revolution*. Boston: Harvard University Press.

MacKenzie, Donald. 2008. *An Engine, Not a Camera: How Financial Models Shape Markets*. Cambridge, MA: MIT Press.

Mamdani, Mahmood. 1996. *Citizen and Subject: Contemporary Africa and the Legacy of Late Colonialism*. Princeton, NJ: Princeton University Press.

Mao, Isaac. "Sharism." Retrieved July 2010, http://www.sharism.org.

———. 2008. "Sharism: A Mind Revolution." In Joi Ito (ed.), *Freesouls*, https://freesouls.cc/essays /07-isaac-mao-sharism.html.

Marcus, George E. 1995. "Ethnography in/of the World System: The Emergence of Multi-Sited Ethnography." *Annual Review of Anthropology* 24: 95–117.

Martin, Emily. 1994. *Flexible Bodies: Tracking Immunity in American Culture—From the Days of Polio to the Age of AIDS*. Boston: Beacon Press.

Marwick, Allison. 2015. "Instafame: Luxury Selfies in the Attention Economy." *Public Culture* 27, no. 1: 137–160.

Massey, Doreen. 1994. *Space, Place, and Gender*. Minneapolis: University of Minnesota Press.

Maxigas. 2012. "Hacklabs and Hackerspaces: Tracing Two Geneologies." *Journal of Peer Production* 5. http://peerproduction.net/issues/issue-2/peer-reviewed-papers/hacklabs-and -hackerspaces/.

McRobbie, Angela. 2010. "Clubs to Companies: Notes on the Decline of Political Culture in Speeded Up Creative Worlds." *Journal of Cultural Studies* 16, no 4: 516–531.

———. 2011. "Reflections on Feminism, Immaterial Labour and the Post-Fordist Regime." *New Formations* 70 (Winter): 60–76.

———. 2016. *Be Creative: Making a Living in the New Culture Industries*. Cambridge: Polity.

Mellis, David A. and Leah Buechley. 2014. "Do-It-Yourself Cellphones: An Investigation into the Possibilities and Limits of High-Tech DIY." In *Proceedings of the ACM SIGCHI Conference on Human Factors in Computing Systems*, April, 1723–1732.

Merkel-Hess, Kate. 2016. *The Rural Modern: Reconstructing the Self and State in Republican China*. Chicago: University of Chicago Press.

Metcalf, Jacob, Emanuel Moss, and danah boyd. 2019. "Owning Ethics: Corporate Logics, Silicon Valley, and the Institutionalization of Ethics." *Social Research: An International Quarterly* 82, no. 2: 449–476.

Miller, Peter and Nikolas Rose. 2008. *Governing the Present*. Cambridge: Polity Press.

Minzner, Carl. 2018. *End of an Era: How China's Authoritarian Revival Is Undermining Its Rise*. Oxford: Oxford University Press.

Mirchandani, Kiran and Winifred Poster. 2016. *Borders in Service: Enactments of Nationhood in Transnational Call Centres. Toronto:* University of Toronto Press.

Mitchell, Timothy. 1991. *Colonising Egypt.* Berkeley: University of California Press.

———. 2002. *The Rule of Experts: Egypt, Techno-politics, Modernity.* Berkeley: University of California Press.

Modrak, Rebekah. 2015. "Learning to Talk Like An Urban Woodsman: An Artistic Intervention." *Journal of Consumption Markets & Culture* 18, no. 6: Communicating Identity/Consuming Difference, 539–558.

Mol, Annemarie. 2002. *The Body Multiple: Ontology in Medical Practice.* Durham, NC: Duke University Press.

Morley, David and Kevin Robins. 1995. "Techno-Orientalism: Japan Panic." *Spaces of Identity: Global Media, Electronic Landscapes and Cultural Boundaries.* New York: Routledge, 147–173.

Mudimbe, Valentin-Yves. 1988. *The Invention of Africa: Gnosis, Philosophy, and the Order of Knowledge.* Bloomington: Indiana University Press.

Mullaney, Thomas S. 2017. *The Chinese Typewriter: A History.* Cambridge, MA: MIT Press.

Murphy, Michelle. 2018. *The Economization of Life.* Durham, NC: Duke University Press.

Nadeem, Shehzad. 2013. *Dead Ringers: How Outsourcing Is Changing the Way Indians Understand Themselves.* Princeton, NJ: Princeton University Press.

Nafus, Dawn. 2012. "'Patches Don't Have Gender': What Is Not Open in Open Source Software." *New Media & Society* 14, no. 4: 669–683.

Nafus, Dawn, James Leach, and Bernhard Krieger. 2006. "Free/Libre and Open Source Software: Policy Support (FLOSSPOLS), Gender: Integrated Report of Findings." University of Cambridge.

Nakamura, Lisa. 2002. *Cybertypes: Race, Ethnicity, and Identity on the Internet.* London: Routledge.

———. 2007. *Digitizing Race: Visual Cultures of the Internet.* Minneapolis: University of Minnesota Press.

———. 2011. "Economies of Digital Production in East Asia: iPhone Girls and the Transnational Circuits of Cool." *Media Fields Journal* 2: 1–10.

———. "Queer Female of Color: The Highest Difficulty Setting There Is? Gaming Rhetoric as Gender Capital." *Ada: A Journal of Gender, New Media, & Technology* no. 1 (November).

Neff, Gina. 2012. *Venture Labor: Work and the Burden of Risk in Innovative Industries.* Cambridge, MA: MIT Press.

Negri, Antonio. 1992. *Marx Beyond Marx: Lessons on the Grundrisse.* London: Pluto Press.

Neilson, Brett and Ned Rossiter. 2008. "Precarity as a Political Concept, or, Fordism as Exception." *Theory, Culture, & Society* 25, nos. 7–8: 51–72.

Nguyen, Josef. 2018. "How Makers and Preppers Converge in Postmodern and Post-Apocalyptic Ruin." *Lateral—Journal of the Cultural Studies Association* no. 7.2.

Norman, Don. 2013. *The Design of Everyday Things.* New York: Basic Books.

Nussbaum, Bruce. 2011. "Design Thinking Is a Failed Experiment." *FastCompany*, May. https://www.fastcompany.com/1663558/design-thinking-is-a-failed-experiment-so-whats-next.

O'Connor, Justin. 2012. "Shanghai Modern: Replaying Futures Past." *Culture Unbound—Journal of Current Cultural Research* 4: 15–34.

O'Donnell, Mary Ann. 2012. "Population and Hukou Update, 2012." *Shenzhen Noted*, July 5. https://shenzhennoted.com/tag/来了就是深圳人/.

———. 2017. "Heroes of the Special Zone: Modeling Reform and Its Limits." In *Learning from Shenzhen: China's Post-Mao Experiment from Special Zone to Model City*, edited by Mary Ann O'Donnell, Winnie Wong, and Jonathan Bach, 39–64. Chicago: University of Chicago Press.

———. 2017. "Laying Siege to the Villages: The Vernacular Geography of Shenzhen." In *Learning from Shenzhen: China's Post-Mao Experiment from Special Zone to Model City*, edited by Mary Ann O'Donnell, Winnie Wong, and Jonathan Bach, 39–64. Chicago: University of Chicago Press.

O'Donnell, Mary Ann, Winnie Wong, and Jonathan Bach, eds. 2017. *Learning from Shenzhen: China's Post-Mao Experiment from Special Zone to Model City.* Chicago: University of Chicago Press.

Oldenziel, Ruth. 1991. *Making Technology Masculine: Men, Women, and Modern Machines in America, 1870–1945.* Amsterdam: Amsterdam University Press.

O'Neil, Cathy. 2016. *Weapons of Math Destruction: How Big Data Increases Inequality and Threatens Democracy.* New York: Crown.

Ong, Aihwa. 1999. *Flexible Citizenship: The Cultural Logics of Transnationality.* Durham, NC: Duke University Press.

———. 2006. *Neoliberalism as Exception: Mutations in Citizenship and Sovereignty.* Durham, NC: Duke University Press.

Ong, Aihwa and Stephen Collier, eds. 2005. *Global Assemblages: Technology, Politics, and Ethics as Anthropological Problems. Oxford, UK:* Blackwell.

Öniş, Ziya. 1991. "Review: The Logic of the Developmental State." *Comparative Politics* 24, no. 1: 109–126.

Oreglia, Elisa. 2014. "ICT and (Personal) Development in Rural China." *Information Technologies & International Development* 10(3): 19–30.

O'Reilly Press Release. 2012. "O'Reilly's Make and Otherlab Win DARPA Mentor Award to Bring Making to Education: Program Will Help Establish Makerspaces in High Schools." January 19, https://www.oreilly.com/pub/pr/2962.

Orlikowski, Wanda J. and Susan V. Scott. 2015. "Exploring Material-Discursive Practices: Comments on Hardy and Thomas' *Discourse in a Material World.*" *Journal of Management Studies* 52, no. 5 (July): 697–705.

Pacific Standard. 2017. "The Future of Work and Workers." Ongoing Series, last updated April 2017. https://psmag.com/tag/the-future-of-work.

Pal, Joyojeet. 2017. "CHI4Good or Good4CHI." In *Extended Abstracts of the ACM SIGCHI Conference on Human Factors in Computing*, Denver, CO, May 6–11.

Pang, Laikwan. 2012. *Creativity and Its Discontents: China's Creative Industries and Intellectual Property Rights Offenses.* Durham, NC: Duke University Press.

Parreñas, Juno Salazar. 2018. *Decolonizing Extinction: The Work of Care in Orangutan Rehabilitation.* Durham, NC: Duke University Press.

Parthasarathy, Shobita. 2017. *Patent Politics: Life Forms, Markets, & The Public Interest in the United States & Europe.* Chicago: University of Chicago Press.

Pasquale, Frank. 2015. *The Black Box Society: The Secret Algorithms That Control Money and Information.* Boston: Harvard University Press.

Peidong, Yang and Tang Lijun. 2018. "'Positive Energy': Hegemonic Intervention and Online Media Discourse in China's Xi Jinping Era." *China: An International Journal* 16, no. 1: 1–22.

Pettis, Bre. 2015. "Why Everyone Is a Maker Now." *Popular Mechanics*, May 28. https://www.popularmechanics.com/home/a15764/bre-pettis-maker-nation/.

Philip, Kavita. 2004. *Civilizing Natures: Race, Resources, and Modernity in Colonial South India.* New Brunswick, NJ: Rutgers University Press.

———. 2005. "What Is a Technological Author? The Pirate Function and Intellectual Property." *Postcolonial Studies* 8, no. 2: 199–218.

Philip, Kavita, Lilly Irani, and Paul Dourish. 2010. "Postcolonial Computing: A Tactical Survey." *Science, Technology, & Human Values* 37, no. 1: 3–29.

Powell, Alison. 2012. "Democratizing Production Through Open Source Knowledge: From Open Software to Open Hardware." *Media, Culture & Society* 34, no. 6: 50–73.

Puig de la Bellacasa, Maria. 2011. "Matters of Care in Technoscience: Assembling Neglected Things." *Social Studies of Science* 41 (1): 85–106.

Qiu, Jack Linchuan. 2009. *Working-Class Network Society: Communication Technology and the Information Have-Less in Urban China*. Cambridge, MA: MIT Press.

———. 2016. *Goodbye iSlave: A Manifesto for Digital Abolition*. Urbana: University of Illinois Press.

———. 2018. "China's Digital Working Class and Circuits of Labor." *Communication and the Public* 3, no. 1: 5–18.

Rabinow, Paul. 1995. *French Modern: Norms and Forms of the Social Environment. Chicago:* University of Chicago Press.

Rankin, Joy Lisi. 2018. *The People's History of Computing in the United States*. Cambridge, MA: Harvard University Press.

Rao, Venkatesh. 2012. "Entrepreneurs Are the New Labor: Part I, II, III." *Forbes*, September. https://www.forbes.com/sites/venkateshrao/2012/09/03/entrepreneurs-are-the-new-labor-part-i/#1023d22c4eab.

Ratto, Matt. 2011. "Critical Making: Conceptual and Material Studies in Technology and Social Life." *Information Society Journal* 27, no. 4, 252–260.

Raval, Noopur and Paul Dourish. 2016. "Standing Out from the Crowd: Emotional Labor, Body Labor, and Temporal Labor in Ridesharing." In *Proceedings of the 19th ACM Conference on Computer-Supported Cooperative Work & Social Commmputing (CSCW'16)*, 97–107.

Rheingold, Howard, ed. 2014. *The Peeragogy Handbook. A Guide for Peer-Learning and Peer Production. Vol. 2*. Arlington, MA: Pierce Press.

Rofel, Lisa. 1999. *Other Modernities: Gendered Yearnings in China After Socialism*. Berkeley: University of California Press.

———. 2007. *Desiring China: Experiments in Neoliberalism, Sexuality, and Public Culture*. Durham, NC: Duke University Press.

———. 2016. "Temporal-Spatial Migration: Workers in Transnational Supply-Chain Factories." In *Ghost Protocol: Development and Displacement in Global China*, edited by Carlos Rojas and Ralph A. Litzinger, 167–190. Durham, NC: Duke University Press.

Roh, David S., Betsy Huang, and Greta A. Niu. 2015. *Techno-Orientalism: Imagining Asia in Speculative Fiction, History, and Media*. New Brunswick, NJ: Rutgers University Press.

Roitman, Janet. 2013. *Anti-Crisis*. Durham, NC: Duke University Press.

Rojas, Carlos and Ralph A. Litzinger, eds. 2016. *Ghost Protocol: Development and Displacement in Global China*. Durham, NC: Duke University Press.

Rose, Nikolas. 1996. "The Death of the Social? Re-Figuring the Territory of Government." *Journal of Economy and Society* 25, no. 3: 327–356.

———. 1998. *Inventing Our Selves: Psychology, Power, and Personhood*. Cambridge: Cambridge University Press.

Rosner, Daniela K. 2018. *Critical Fabulations: Reworking the Methods and Margins of Design*. Cambridge, MA: MIT Press.

Rosner, Daniela K., Samantha Shorey, Brock Craft, and Helen Remick. 2018. "Making Core Memory: Design Inquiry into Gendered Legacies of Engineering and Craftwork." In *Proceedings of the 2018 CHI Conference on Human Factors in Computing Systems (CHI'18)*, Paper No. 531.

Ross, Andrew 2004. *No Collar: The Humane Workplace and Its Hidden Costs*. Philadelphia: Temple University Press.

———. 2007. *Fast Boat to China: High-Tech Outsourcing and the Consequences of Free Trade—Lessons from Shanghai*. New York: Vintage Books.

Rossiter, Ned. 2016. *Software, Infrastructure, Labor: A Media Theory of Logistical Nightmares*. London: Routledge.

Roush, Wade. 2012. "HAXLR8R Startups Report Back from Shenzhen, the Hardware Candyland." *Xconomy*, June 20. https://xconomy.com/san-francisco/2012/06/20/haxlr8r-startups-report-back-from-shenzhen-the-hardware-candyland/.

Roy, Ananya. 2010. *Poverty Capital: Microfinance and the Making of Development*. London: Routledge.

Russo, Gina. 2009. "Better City, Better Life." *China Beat Blog Archive 2008–2012*, 527. http://digitalcommons.unl.edu/chinabeatarchive/527.

Sacks, David O. and Peter Thiel. 1998. *The Diversity Myth: Multiculturalism and Political Intolerance on Campus*. Oakland, CA: Independence Institute.

Said, Edward. 1978. *Orientalism*. New York: Pantheon Books.

Sandvig, Christian. 2012. "Connection at Ewiiaapaayp Mountain: Indigenous Internet Infrastructure." In *Race After the Internet*, edited by Lisa Nakamura and Peter Chow-White, 168–200. New York: Routledge.

———. 2019. *Corrupt Personalization: Algorithmic Discrimination in Media and Information*. Draft, April 4.

Sanval, Kalyan. 2007. *Rethinking Capitalist Development: Primitive Accumulation, Governmentality, & Post-Colonial Capitalism*. London: Routledge.

Saxenian, AnnaLee. 1996. *Regional Advantage: Culture and Competition in Silicon Valley and Route 128*. Cambridge, MA: Harvard University Press.

Schein, Louisa. 1997. "Gender and Internal Orientalism in China." *Modern China* 23, no. 1 (January): 69–98.

———. 1999. "Performing Modernity." *Cultural Anthropology* 14, no. 4 (August): 361–395.

———. 2000. *Minority Rules: The Miao and the Feminine in China's Cultural Politics*. Durham, NC: Duke University Press.

Scholz, Trebor, ed. 2012. *Digital Labor: The Internet as Playground and Factory*. London: Routledge.

———. 2016. *Uberworked and Underpaid: How Workers Are Disrupting the Digital Economy*. New York: Polity.

Scholz, Trebor and Nathan Schneider. 2017. *Ours to Hack and to Own. The Rise of Platform Cooperativism, a New Vision for the Future of Work and a Fairer Internet*. New York: OR Books.

Scott, James C. 1987. *Weapons of the Weak: Everyday Forms of Peasant Resistance*. New Haven, CT: Yale University Press.

———. 1998. *Seeing Like a State: How Certain Schemes to Improve the Human Condition Have Failed*. New Haven, CT: Yale University Press.

Sengers, Phoebe, Kirsten Boehner, Shay David, and Joseph 'Jofish' Kaye. 2005. "Reflective Design." In *Proceedings of the 5th Decennial Conference on Critical Computing, Aarhus: Between Sense and Sensibility*, 49–58.

Shambaugh, David. 2016. *China's Future*. Cambridge, UK: Polity.

Shen, Hsui-Hua. 2008. "The Purchase of Transnational Intimacy: Women's Bodies, Transnational Masculine Privileges in Chinese Economic Zones." *Asian Studies Review* 32, no. 1: 57–75.

Shirky, Clay. 2008. *Here Comes Everybody: The Power of Organizing Without Organizations*. New York: Penguin.

———. 2010. *Cognitive Surplus: Creativity and Generosity in a Connected Age*. New York: Penguin.

———. 2015. *Little Rice: Smartphones, Xiaomi, and the Chinese Dream*. New York: Columbia Global Reports.

Silberman, Michael "Six," Lilly Irani, and Joel Ross. 2010. "Ethics and Tactics of Professional Crowdwork." In *XRDS: Crossroads, The ACM Magazine for Students*, 17, no. 2: 39–43.

Simon, Scott. 2018. "Driver's Suicide Highlights 'Race to the Bottom' in Cab Industry, Union Director Says." *NPR*, February 10. https://www.npr.org/2018/02/10/584757778/taxi-drivers-face-financial-crisis.

Simonite, Tom. 2016. "Moore's Law Is Dead. Now What?" *MIT Technological Review*. May 13. https://www.technologyreview.com/s/601441/moores-law-is-dead-now-what/.

Sims, Christo. 2017. *Disruptive Fixation: School Reform and the Pitfalls of Technoidealism*. Princeton, NJ: Princeton University Press.

Sivek, Susan Currie. 2011. "'We Need a Showing of Hands': Technological Utopianism in MAKE Magazine." *Journal of Communication Inquiry* 25, no. 3: 187–209.

Söderberg, Johan and Alessandro Delfanti. 2015. "Hacking Hacked! The Life Cycles of Digital Innovation." *Science, Technology, & Human Values* 40, no. 5: 793–798.

Solinger, Dorothy. 1999. *Contesting Citizenship in Urban China: Peasant Migrants, the State, and the Logic of the Market*. Berkeley: University of California Press.

Song, Geng and Derek Hird. 2014. *Men and Masculinity in Contemporary China*. Leiden: Brill.

Spivak, Gayatri Chakravorty. 1999. *A Critique of Postcolonial Reason: Toward a History of the Vanishing Present*. Boston: Harvard University Press.

Srnicek, Nick. 2016. *Platform Capitalism*. Cambridge: Polity.

SSL Nagbot. 2016. "Feminist Hacking/Making: Exploring New Gender Horizons of Possibility." *Journal of Peer Production* 8. http://peerproduction.net/issues/issue-8-feminism-and-unhacking-2/feminist-hackingmaking-exploring-new-gender-horizons-of-possibility/.

Steffan, Philip. 2015. "Arduino gegen Arduino: Gründer streiten um die Firma." *Heise Online*. February 15. https://www.heise.de/make/meldung/Arduino-gegen-Arduino-Gruender-streiten-um-die-Firma-2549653.html.

Stengers, Isabelle and Vinciane Despret, translated byApril Knutson. 2014. *Women Who Make a Fuss: The Unfaithful Daughters of Virginia Woolf*. Minneapolis, MN: Univocal.

Stoler, Ann Laura. 1995. *Race and the Education of Desire: Foucault's History of Sexuality and the Colonial Order of Things*. Durham, NC: Duke University Press.

———. 2016. *Duress: Imperial Durabilities in Our Times*. Durham, NC: Duke University Press.

Stone, Madeline. 2014. "Tencent Is Building a Stunning New 'Vertical Campus' in China." *Business Insider*. March 18. https://www.businessinsider.com/new-tencent-building-2014-3.

Suchman, Lucy. 2006. *Human-Machine Reconfiguration: Plans and Situated Action,* 2nd ed. Cambridge: Cambridge University Press.

———. 2011. "Anthropological Relocations and the Limits of Design." *Annual Review of Anthropology* 40: 1–18.

Suchman, Lucy, Randall Trigg, and Jeanette Blomberg. 2002. "Working Artefacts: Ethnomethods of the Prototype." *British Journal of Sociology* 53, no. 2: 163–179.

Sunder Rajan, Kaushik, ed. 2012. *Lively Capital: Biotechnologies, Ethics, and Governance in Global Markets*. Durham, NC: Duke University Press.

Tadiar, Neferti X. M. 2009. *Things Fall Away: Phillipine Historical Experience and the Making of Globalization*. Durham, NC: Duke University Press.

———. 2013. "Life-Times of Disposability Within Global Neoliberalism." *Social Text* 115, vol. 31, no. 2: 19–48.

Takheteyev, Yuri. 2012. *Coding Places: Software Practice in a South American City*. Cambridge, MA: MIT Press.

Tan, Christopher T. 2015. "Techno-Orientalism with Chinese Characteristics: Maureen F. McHugh's China Mountain Zhang." *Journal of Transnational American Studies* 6, no. 1 (January).

Taylor, Alex S. 2011. "Out There." In *Proceedings of the SIGCHI Conference on Human Factors in Computing Systems (CHI 2011)*, 685–694.

Terranova, Tiziana. 2000. "Free Labor: Producing Culture for the Digital Economy." *Social Text* 18, no. 2: 33–58.

The State Council. 2015a. "China Encourages Innovation." The People's Republic of China. February 7. http://english.gov.cn/news/video/2015/02/07/content_281475051861119.htm.

The State Council. 2015b. "Premier Li Keqiang Visits Makerspace in Shenzhen." The People's Republic of China. January 4, 2015. http://english.www.gov.cn/premier/photos/2015/01/04/content_281475034064167.htm.

The State Council. 2016. "Full Transcript of the State Council policy briefing on Feb 5, 2016." The People's Republic of China. http://english.www.gov.cn/news/policy_briefings/2016/02/05/content_281475284749774.htm.

The State Council. 2017. "China to Establish Made in China 2025 National Demonstration Zones." The People's Republic of China. November 23. http://english.gov.cn/policies/latest_releases/2017/11/23/content_281475952054656.htm.

Thiel, Peter and Blake Masters. 2014. *Zero to One: Notes on Startups, or How to Build the Future*. New York: Crown Business.

Thompson, Laurence G. 2005. *Ta T'ung Shu: The One-world Philosophy of K'ang Yu-Wei (1958)*. Abingdon, UK and New York: Routledge.

Tolentino, Jia. 2017. "The Gig Economy Celebrates Working Yourself to Death." *New Yorker*, March 22. https://www.newyorker.com/culture/jia-tolentino/the-gig-economy-celebrates-working-yourself-to-death.

Traweek, Sharon. 1988. *Beamtimes and Lifetimes: The World of High Energy Physicists*. Cambridge, MA: Harvard University Press.

Tronti, Mario. 1966. *Operai e Capitale*. Torino: Einaudi.

Tsing, Anna Lowenhaupt. 2000. "Inside the Economy of Appearance." *Public Culture* 12, no. 1: 115–144.

———. 2004. *Friction: An Ethnography of Global Connection*. Princeton, NJ: Princeton University Press.

———. 2015. *The Mushroom at the End of the World: On the Possibility of Life in Capitalist Ruins*. Princeton, NJ: Princeton University Press.

Turkle, Sherry. 2005. *The Second Self: Computers and the Human Spirit*. Cambridge, MA: MIT Press.

Turner, Fred. 2006. *From Counterculture to Cyberculture: Stexwart Brand, The Whole Earth Network, and the Rise of Digital Utopianism*. Chicago: University of Chicago Press.

———. 2013. *The Democratic Surround: Multimedia and American Liberalism from World War II to the Psychedlic Sixties*. Chicago: University of Chicago Press.

———. 2018. "Millenarian Tinkering: The Puritan Roots of the Maker Movement." *Technology and Culture* 59, no. 4: 160–182.

Tushnet, Rebecca. 2004. "Copy this Essay: How Fair Use Doctrine Harms Free Speech and How Copying Serves It." *Yale Law Journal* 114: 535–587.

Ueno, Toshiya. 1999. "Techno-Orientalism and Media-Tribalism: On Japanese Animation and Rave Culture." *Third Text* 13: 47, 95–106.

Vinsel, Lee. 2017. "Design Thinking Is Kind of Like Syphilis—It's Contagious and Rots Your Brains." *Medium*, December.

Vogel, Eric. 2017. "Forward." In *Learning from Shenzhen: China's Post-Mao Experiment from Special Zone to Model City*, edited by Mary Ann O'Donnell, Winnie Wong, and Jonathan Bach, vii–xiv. Chicago: University of Chicago Press.

Vogel, Ezra F. 2011. *Deng Xiaoping and the Transformation of China*. Cambridge, MA: Harvard University Press.

Von Hippel, Eric. 2005. *Democratizing Innovation*. Cambridge, MA: MIT Press.

von Schnitzler, Antina. 2016. *Democracy's Infrastructure: Techno-Politics and Protest after Apartheid*. Princeton, NJ: Princeton University Press.

Vora, Kalindi. 2015. *Life Support: Biocapital and the New History of Outsourced Labor*. Minneapolis: University of Minnesota Press.

Wajcman, Judy. 1991. *Feminism Confronts Technology*. Cambridge: Polity.

———. 2016. *Pressed for Time: The Acceleration of Life in Digital Capitalism*. Chicago: University of Chicago Press.

Wajcman, Judy and Nigel Dodd, eds. 2017. *The Sociology of Speed: Digital, Organizational, and Social Temporalities*. Oxford: Oxford University Press.

Wallis, Cara. 2013a. *Technomobility in China: Young Migrant Women and Mobile Phones*. New York: New York University Press.

———. 2013b. "Technology and/as Governmentality: The Production of Young Rural Women as Low-Tech Laboring Subjects in China." *Communication and Critical/Cultural Studies* 10, no. 4: 341–358.

Wallis, Cara and Jack Linchuan Qiu. 2009. "*Shanzaiji* and the Transformation of the Local Mediacape in Shenzhen." In *Mapping Media in China: Region, Province, Locality*, edited by Wanning Sun and Jenny Chio, 109–125. London: Routledge.

Warner, Michael. 2005. *Publics and Counterpublics*. New York: Zone Books.

Wasserstrom, Jeffrey. 2009. *Global Shanghai, 1850–2010: A History of Fragments*. London: Routledge.

Wasserstrom, Jeffrey and Elizabeth Perry. 2019. *Popular Protest and Political Culture in Modern China*, 2nd ed. London: Routledge.

Webster, Frank. 2006. *Theories of the Information Society*. London: Routledge.

Whittaker, Meredith, Kate Crawford, Roel Dobbe, Genevieve Fried, Elizabeth Kaziunas, Varoon Mathur, Sarah Myers West, Rashida Richardson, Jason Schultz, and Oscar Schwartz. 2018. "AI Now Report 2018." December. https://ainowinstitute.org/AI_Now_2018_Report.pdf.

Whyte, Martin King and William L. Parish. 1984. *Urban Life in Contemporary China*. Chicago: University of Chicago Press.

Whyte, Martin King. 2010. *Myth of the Social Volcano: Perceptions of Inequality and Distributive Injustice in Contemporary China*. Stanford, CA: Stanford University Press.

———. 2012. "China's Post-Socialist Inequality." *Current History*, September, pp. 229–234.

Wielander, Gerda and Derek Hird, eds. 2018. *Chinese Discourses on Happiness*. Hong Kong: Hong Kong University Press.

Wilcox, Emily. 2012. "Selling Out Post-Mao: Dance Work and Ethics of Fulfullment in Reform Era China." In *Chinese Modernity and the Individual Psyche*, edited by Andrew Kipnis, 43–65. New York: Palgrave Macmillan.

———. 2018. *Revolutionary Bodies: Chinese Dance and the Socialist Legacy*. Oakland: University of California Press.

Williams, Elliot. 2015. "Arduino v. Arduino." *Hackaday*, February 25. https://hackaday.com/2015/02/25/arduino-v-arduino/.

Wong, Winnie. 2013. *Van Gogh on Demand: China and the Readymade*. Chicago: University of Chicago Press.

Wübbeke, Jost, Mirjam Meissner, Max J. Zenglein, Jaqueline Ives, and Björn Conrad. 2016. *Made in China 2025: The Making of a High-Tech Superpower and Consequences for Industrial Countries*. Berlin: Merics: Mercator Institute for China Studies.

Wuwei, Li. 2011. *How Creativity Is Changing China*, edited by Michael Keane. London: Bloomsbury.

Xiang, Biao. 2016. "'You've Got to Rely on Yourself . . . and the State!': A Structural Chasm in the Chinese Political Moral Order." In *Ghost Protocol: Development and Displacement in Global China*, edited by Carlos Rojas and Ralph A. Litzinger, 131–149. Durham, NC: Duke University Press.

Xinhua News. 2015. "李克强鼓励"创客"小伙伴: 众人拾柴火焰高." November 11. http://www.xinhuanet.com//politics/2015-01/11/c_127376994.htm.

Yan, Hairong. 2008. *New Masters, New Servants: Migration, Development, and Women Workers in China*. Durham, NC: Duke University Press.

Yang, Fan. 2016a. *Faked in China: Nation Branding, Counterfeit Culture, and Globalization*. Bloomington: Indiana University Press.

———. 2016b. "Fiscal Orientalism: China Panic, the Indebted Citizen, and the Spectacle of National Debt." *Journal of Asian American Studies* 19, no. 3 (October): 375–396.

Yang, Guobin. 2009. *The Power of the Internet in China: Citizen Activism Online*. New York: Columbia University Press.

———. 2017. *The Red Guard Generation and Political Activism in China*. New York: Columbia University Press.

Yang, Jie, ed. 2014a. *The Political Economy of Affect and Emotion in East Asia*. New York: Routledge.

———. 2014b. "The Happiness of the Marginalized: Affect, Counseling, and Self-Reflexivity in China." In *The Political Economy of Affect and Emotion in East Asia*, edited by Jie Yang, 45–62. New York: Routledge.

Zaloom, Caitlin. 2010. *Out of the Pits: Traders and Technology from Chicago to London*. Chicago: University of Chicago Press.

Zhan, Mei. 2009. *Other-Worldly: Making Chinese Medicine through Transnational Frames*. Durham, NC: Duke University Press.

Zhang, Li. 2008. "Private Homes, Distinct Lifestyles: Performing a New Middle Class." In *Privatizing China: Socialism from Afar*, edited by Li Zhang and Aihwa Ong, 23–40. Ithaca, NY: Cornell University Press.

Zhang, Li, and Aihwa Ong. 2008. *Privatizing China: Socialism from Afar*. Ithaca, NY: Cornell University Press.

Zhang, Lin and Anthony Fung. 2013. "The Myth of 'Shanzhai' Culture and the Paradox of Digital Democracy in China." *Inter-Asia Cultural Studies* 14, no. 3: 401–416.

Zhang, Lu. 2019. "A Race to the Bottom or Variegated Labor Regimes? Capital Monopoly and Labor Politics in China's Electronics Industry." Paper presented at the 2019 Association for Asian Studies (AAS), Denver, CO.

Žižek, Slavoj. 2011. *Living in the End Times*. London: Verso.

Zuboff, Shoshana. 2019. *The Age of Surveillance Capitalism: The Fight for a Human Future at the New Frontier of Power*. New York: PublicAffairs.

INDEX

Note: Page numbers in italic type indicate illustrations.

aaajiao, 48, 53, 66
accelerator programs, 13, 119, 120, 154, 156, 195
Acer, 197
Adams, Christopher, 59
affect: CCP's use of, 18–19, 203; of change, 170; for Chinese technological innovation, 62; of citizens for China, 19, 25, 140; for counterculture, 42, 68–69, 73, 223; for economization, 13, 18–19, 228n54; for experimentation, 84; for finance capitalism, 134; governance through, 84; linked to prototypes, 2; neoliberalism's use of, 129; openness and participation as ideals productive of, 43, 228n54; for self-economization, 73; for Shenzhen, 106; between start-up and factory, 160–61; as tool of market devices, 14
affect of intervention, 2, 11, 13, 43, 57, 137, 215, 216
agency. *See* affect of intervention; intervention; resistance
Ahmed, Sara, 1, 13, 70, 153, 174–75, 203, 213, 223, 225n5
AI. *See* artificial intelligence
alternatives: to conventional ways of living, 1–2, 6, 11, 22, 25, 46, 48–49, 53, 56–57, 70–72, 76–77, 147; technology as means of discovering/exploring, 1–4, 12, 217–19; yearnings for, 14, 24, 30–31, 37, 83, 90, 134, 147, 151, 158, 213, 215, 221, 223
Altman, Mitch, 67–69, 69
Amazon, 14, 146, 150, 167
American Tax Cuts and Job Act, 215
Amrute, Sareeta, 231n118, 231n119
Anagnost, Ann, 46, 233n17
Anderson, Chris, 12, 119, 122, 124, 154
anticipation, 12, 129, 133
Apple, 85, 93–95, 108, 113, 149, 172, 178, 197
apprenticeship model, 190, 194

Arduino, 7, 53, 66, 67, 91, 95–96, 106–12, 116, 178, 238n93
Arduino SRL, 109
Ars Electronica, 60, 62
art education, 49
artificial intelligence (AI), 10, 191, 222
Asian tiger economies, 81
Association for Computing Machinery, 219
Athisaari, Marko, 60
aura, 113, 115
Autodesk, 137–38
automation machinery, 187
autonomous Marxism, 146, 149, 241n11
autonomy, 9–12, 46
AVIC, 183

Bach, Jonathan, 26, 27
backwardness: China associated with, 2, 4, 19, 37, 41, 63, 76–78, 86–88, 122, 137, 148, 150–51, 158, 214; migrant workers associated with, 150; opportunity associated with, 4, 37, 76–77, 116, 139, 148; Shenzhen associated with, 15, 116, 139
Baishizhou, China, 27, 28
Banzi, Massimo, 107–11, 107, 178
BarCamps, 41, 44, 65–66, 124
Barragan, Hernando, 112
Bauhaus, 14, 108–9
Baym, Nancy, 61, 161
Beijing Maker Carnival, 67, 71, 72
Bell, Genevieve, 157
Belt and Road initiative (BRI), 23–24, 175–76, 220
Benjamin, Ruha, 226n17
Benjamin, Walter, 113, 115
Benkler, Yochai, 60
Berkman Klein Center, 46, 59, 62
Berlant, Lauren, 80, 226n17
Bhabha, Homi, 70

A NOTE ON THE TYPE

This book has been composed in Adobe Text and Gotham.
Adobe Text, designed by Robert Slimbach for Adobe,
bridges the gap between fifteenth- and sixteenth-century
calligraphic and eighteenth-century Modern styles.
Gotham, inspired by New York street signs, was designed
by Tobias Frere-Jones for Hoefler & Co.

Printed and bound by CPI Group (UK) Ltd, Croydon, CR0 4YY

27/10/2024

14580234-0001